Apollo 14

Mutiny

David Noever

The first mission to the Moon after the disastrous events of Apollo 13.

The return to spaceflight for the first American in space.

A career-ending bout with a pilot's worst nightmare, uncontrollable dizzy spells.

An unsanctioned science experiment to test if human extrasensory perception could reach to the Moon and back.

An often rehearsed space docking that succeeded every time except when it mattered the most.

When the astronauts blinked their eyes, floating lights accompanied them as silent escorts.

An often rehearsed lunar landing seemed impossible if radar could not tell them what craters and boulders they were blind to below them.

A failure was inconceivable if Apollo was ever to return to the Moon.

And when failure came, the answer was mutiny.

Contents

Chapter 1

Part 1: Senate

The man in the white vest was Eugene Kranz, the iconic leader of Houston's Mission Control. He was in an alien environment. Kranz was making his way into the US Senate hearing on an Apollo investigation into the third attempt to land men on the moon. That mission, Apollo 13, had set the bar. It was the first failed attempt to step foot on lunar soil. It had proceeded flawlessly until an explosion had crippled the flight too far to return and too near not to continue a daring slingshot roundtrip around the moon and back home. And Gene Kranz was at the center of the maelstrom.

Senator Charles Percy of Illinois shook hands with Eugene Kranz, who took his seat in the Senate hearings. Percy had a movie-star quality with chiseled features. He was exploring a future run for President. Before his political career, Percy had led the Chicago-based *Bell and Howell Company*, a pioneer in making movie cameras for home and military use. All the press cameras now focused on the first political-space collision.

"So you're the guy with the white vest? I hear you got a photographic memory," said Percy, momentarily impressed with Kranz.

"Sorry, I didn't catch the name," said Kranz.

"Senator Percy," he said then paused with a sour face. "From a state far away from Texas."

Senator Walter Mondale is reading a four inch stack of reports called the "Tiger Team Reports." The report sits hidden in his lap under the table. Mondale, a Minnesota Democrat, had been offered the Vice Presidency in 1972 and in a rare political revolt, declined the offer to run.

"Mr. Kranz," began Mondale, "you would agree with the President about the enormity of these events?" Mondale flipped a page of the report held in his lap under the table.

NASA Flight Controller, Eugene Kranz, is a determined, bull-dog of a man, who did not mince words with his direct answers. "I believe the President was moved by the moment. We all were," said Kranz then paused. "But, as I recall, on the aircraft carrier, he said the moon landing was the greatest single human event since..." Kranz pauses, somewhat embarrassed. "Well, even Nixon said it, since the Creation."

"As you know then, the Tiger Team describes the nature of the Apollo 13 accident," said Mondale.

"Yes, the oxygen storage tank had been dropped during manufacture," said Kranz. "It had knocked a tube loose."

Mondale closed the report in his lap. "Dropped?"

"Yes Senator. It fell on the floor," said Kranz. "I believe about 2 inches."

"Two inches?" Mondale replied with disbelief. "That's why we're here?"

"In my business, big mistakes are often the result of many tiny ones," said Kranz. "That oxygen tank was originally delivered for the Apollo 10 flight."

"So I'd be correct saying this whole tragedy could've preempted the first moon landing?" said Mondale. "On Apollo 11?"

"We never would have made it to number 13. That 2 inches would have kept us short," said Kranz. "About a quarter million miles short of the moon."

"So your readiness to…" said Mondale.

Kranz interrupted. "I might add, Senator. If we hadn't lost the three astronauts in the Apollo 1 fire and all those corrective safety and insulation changes we certainly would have lost Apollo 13. There was so much water on the panels when 13 tried to restart an electrical short would have sparked an internal fire," said Kranz with a somber tone. "It was a great sacrifice of 3 men. To save 3 men."

"Exactly my point," said Senator Percy. "If men's lives are at stake then there we may have a criminal wrong here."

Mondale continued flipping pages of the Tiger Team report. "Now the actual explosion, itself. It happened on the outbound trip to the moon. Can you tell this committee what would have happened if the explosion had come at another time?"

"You mean later or earlier?" asked Kranz.

"Either," replied Mondale.

Kranz recounted from memory, "An explosion of this kind on the launch pad?" Kranz thought with the precision of an engineer and the directness of a medical examiner. "Unless the rescue tower was still primed, it would be fatal."

Mondale leaned back in his chair. "Launch, it's the most dangerous?" Mondale put his hands behind his head and considered his inquisitor's role on the Congressional Aeronautical and Space Sciences Committee. Mondale had become the Senate's attack lead when it came to how NASA policed its contractors. His role focused particularly on North American Aviation following the Apollo 1 launch pad fire that killed Gus Grissom, Ed White and Roger Chaffee in 1967.

"Highest risk, yes," said Kranz. "Later, if the men are already in earth orbit, such an explosion is recoverable. We could call a quick reentry and splashdown."

"So, in the mission plan on the outward leg of the moon trip, it was the worst place?" Mondale added, "On the 13th of April?"

"At that moment, as I recall, the men were just out of our reach," said Kranz. "The moon's gravity was quickly pulling them away from us. We could have never caught them. So we rigged a lunar boomerang."

"Any other critical mission points?" asked Mondale.

"Two," answered Kranz. "If the explosion came one day later, when the three astronauts are circling in lunar orbit, we wouldn't have had enough power to free them from the moon."

"Forever?" asked Mondale.

"Permanent lunar orbit, yes sir," said Kranz.

"And a day later," continued Mondale, "if the explosion had come while the other two astronauts were walking on the moon?"

"Fatal for everyone," said Kranz. "Asphyxia in 15 minutes for the pilot."

"But the moonwalkers would be free...," interjected Mondale.

"From the moon, they would require the mother-ship to get back," said Kranz.

"So this accident, if it comes at any other point in the mission, even slipping a day, then you..." continued Mondale. "You would have had to call in the President?"

"In that moment, yes," said Kranz. He paused. "For the last goodbyes to the families."

Part 2: NASA Astronaut Office

Astronaut Chief Deke Slayton was holding court among the astronauts who heeded his words. The NASA astronaut office appeared like many NASA offices, some cross between an elementary school without windows and a Soviet-style monolith. But this particular office had the buzz of a forthcoming mission to the moon. Coughs and chairs scooting along the floor signaled that the meeting was about to commence. The participants were pilots, mostly from the Navy and mostly those specialized in night landing of planes onto aircraft carriers in the middle of the vast Pacific Ocean.

"Gentlemen, can I have your undivided attention?" said Slayton although not really asking a question. Sounds of rustling movements among his audience of astronauts continued.

"This is Apollo 14," said Slayton. "That follows 13. You understand me."

The room became still.

"Any of you space jockeys here afraid of falling off that horse?" asked Slayton.

"No sir." The cacophony of background noise was a consensus of men who competed for the chance to outpace the man seated next to them. "Not me. Let's go."

"Well then..," said Slayton. He paused. "You're better men than me."

Surprise and silence ensued. Slayton was a mountain of stability and courage, with a chiseled face and a determined, hard-set chin.

"I'm just a man and a pilot and there's only one thing tougher than taking a bad fall from a horse, and that's to get back on the same horse," said Slayton.

Slayton paused. "So be on notice. I'm here to tell you, honestly and simply, you're not on this team-the team of people going back to the moon- unless, one..." Slayton held up one finger to begin a list of prerequisites. "...you're afraid of falling off that horse. Yes sir I am. And two..." Slayton held up a second finger. "....more importantly, you're more afraid now than ever before because once you've been thrown, it's tougher to dust yourself off and get back on that same horse."

The raucous room of pilots who cheered 'let's go' now was silent. Slayton continued "Am I clear?"

From the back, Deke Slayton's head was more square than rounded. He had little patience for technical reasons not to do something that he had set his mind to. But the room in front of Slayton included technicians, astronauts and the soon to be named, Apollo 14 astronauts.

Alan Shepard, the first American in space, was called "The Fearless Leader" by one of his two rookie pilots, Stuart Roosa. The other rookie, Edgar Mitchell, was known by all in the astronaut corps as "The Brain".

"Good. Now let's go to the moon," said Slayton.

As Slayton lifted his palm off of the Apollo 14 flight plan in his hands, a wet palm print was left behind from his sweat.

Alan Shepard got up slowly and steadied himself from falling with a hand on his desk. He turned his neck slowly to the side to adjust himself.

As the room emptied, one astronaut remains, Stuart Roosa. Slayton began to gather up his papers to leave.

"You believe in that 13 bad luck?" asked Roosa.

"Oh it was definitely luck," said Slayton.

"So what now? We're all running out of tokens," said Roosa.

"I mean, *good luck*. Apollo 13 was lucky as hell. That big explosion did not destroy the machine," said Slayton. "It did not penetrate the Command Module and did not destroy the pre-return trajectory. That is just flat, unadulterated luck."

Slayton had gathered his papers and was heading to the door. "So no way anybody's going to take Fourteen from us. Not now."

"In less than nine months, we've lost our shot at Apollo 18, 19, and 20," said Roosa. "That's a helluva' Congressional bust rate." It was both relevant and strange to hear trained pilots talk about the budgetary fights going on within NASA.

"Listen, after the Apollo 1 fire, I cried. They couldn't even get the spacesuits off my friends," said Slayton. He looked at the paper stack in his hands and paused at the door. "The damn plastic melted onto them."

"So on my desk, now I've got a signed picture," said Slayton. "Their last photo. Three smiling faces saying: *It's not that we don't trust you Deke, but this time we decided to go over your head*." Slayton stopped himself with the somber pause of someone who bore a weight without explanation.

Slayton caught Roosa's eyes directly, "Puts a damn lump in your throat. So don't worry, we got your ass covered."

"If something's going to go haywire...," said Roosa.

Slayton interrupted, "...if Apollo 14 does, you'd better buy a comfortable backyard chair and a powerful telescope. Because after Fourteen, your backyard is as close as anybody's getting to the stars."

Edgar Mitchell had a head filled with millions of operational details but could still register the singularity of a call to act now. Mitchell pulled Roosa from the face-to-face questions with the head of the Astronaut Office. "No use getting speculative with old Deke. Shepard and him, they were the Original Seven. No reason to speculate when you've been there," said Mitchell.

Roosa tried to explain his questions: "Shepard, OK. First American in space. Ice in his blood. But Deke?"

"He's curing himself, you know," interrupted Mitchell.

"Slayton's been grounded for twenty years. It's like all great swimming teams have to have a coach who won't go near the water," said Roosa.

"The guy's curing himself," said Mitchell. "Both him and Shepard. First Slayton's heart and then Shepard's ears. That's why you don't have to speculate with these guys. Why speculate about something, when you can just go out and do it?"

"You can't just cure a heart murmur," said Roosa.

Mitchell had a turn to him that could cut through details. "I'm telling you, the guys done it! Real alphabet soup. Took enough different kinds of vitamins to start a pharmacy," said Mitchell.

"Vitamins?" asked Roosa incredulously. "Flight school taught that vitamins just made for expensive urine."

"I'm telling you, these old astronauts, they're a breed," said Mitchell.

Roosa opens the door and Mitchell stands in doorway with the last phrase. "So you tell me this, you're Deke Slayton, every morning you wouldn't drink an extra glass of orange juice? If it meant you could go to the moon?"

Part 3: Dizziness

Astronaut Tom Stafford met Alan Shepard privately that night at a Houston restaurant. Their conversation began in a confidential hush, like two doctors conferring outside a hospital waiting room.

Stafford leaned in towards Shepard, "Al, maybe I've got something for you."

A waitress brought to the table two tall drinks.

"Double," said Shepard signaling to the waitress for a stronger elixir.

"Aren't you some of those spacemen?" said the waitress in her south Texas accent.

Shepard pointed to Slayton, his boss and medical conspirator. "He is. I'm just a desk man," said Shepard.

The waitress knew when casual conversation wasn't welcomed and muttered as she walked away, "A thirsty deskman."

Stafford handed Shepard a napkin with a name scribbled on it. "This is what I've got for you. The guys were talking about some doctor in Los Angeles who's pulling off cures for problems other M.D.'s say can't be cured. They say he's a specialist in ear, nose and throat. Nothing unusual about that, but the word is that he's developed a surgical procedure. He can get rid of this ear problem you've got. Get your flight status back."

"Surgical?" said Shepard. His problem had become chronic and the symptoms had all the features of a curse to a pilot. Shepard had vertigo. If he moved his head the wrong way, the room started to spin the other way. It was an ailment that was taking him off any flight rotations, including a trip to the moon. And Alan Shepard had been America's first man in space.

"Uh huh," answered Stafford. "You're being treated now by medication, right?" Shepard nodded. "Doing any good?"

"I'm going deaf now in my left ear," said Shepard. He looked curiously at the napkin on the table. "You said he uses surgery?"

"That's what I'm told. More than that, whatever he does, it works. You've got nothing to lose. Why don't you go see the man?"

Part 4: Meniere's Syndrome

Alan Shepard caught a plane to visit the Los Angeles doctor, Dr. William House. To conceal his trip, he travelled under an assumed name.

"The head X-rays are back," said Dr. House.

"Transparent, I hope," said Shepard. He was uncomfortable in the presence of anyone authorized either to carry a stethoscope or capable of keeping him out of the sky.

"In my judgment, Mr. Shepard, you're what we call a classic Meniere's case," said House.

"Which means I fall down a lot," said Shepard dryly. "Can you help me?"

"Bright lights bother you?" said House.

"Sure, when an attack starts coming, I don't drive--day or night," said Shepard.

"Ears ringing now?" said House.

"Something like a dog whistle in my head," said Shepard.

"Any sweating or vomiting?" continued House working through his notes.

"I don't even tell the NASA doctors that," said Shepard. "A pilot saves his vomit stories for his priest and his mother."

"I'd recommend putting some handrails near spots in your house-- you know, the bathroom, kitchen, where ever you may have to get up quickly," said House.

Shepard stood to leave and steadied himself on the desk. "I hoped we were clear. This is not about getting up from the dinner table," said Shepard. "I've got only one UP for me to go to."

"The standard treatment for Meniere's is drugs. Diuretics will dry you up like a raisin. Then antihistamines will relax the ear canals."

"It's not likely antihistamines are going to help me operate heavy machinery." Shepard turned to leave. "To the moon."

"Sit down, please. We can always get you under the knife," said House.

"Nobody's going to…" interrupted Shepard.

"In some rare cases, relief is given by surgical destruction of the affected ear. In your case, first I'll try to drain the fluid."

"Uh, drain?" said Shepard, his interest piqued.

House gestured at a diagram as he speaks. "I'd surgically cut a hole in your fluid sack. Insert a tube from the ear canal to a gland, just here, that one leads into the spinal column."

"The spine?" said Shepard, wincing.

"That will drain off some of the fluid, just for starters."

"You're going to cut my backbone?" said Shepard. He stood up very steady to leave.

"One last point, Mr. Shepard."

Shepard paused in his exit.

"Most Meniere's patients have to avoid ladders. Permanently."

"Then you've got to make me a non-patient," said Shepard. "This country has one ladder to climb. I happen to have 9 rungs of that ladder."

"I was thinking more about ladders for painting the house."

"You get me up and down that moon ladder without throwing up and you've got it," said Shepard.

"Hell, Al, if we can put a man on the moon then surely I can silence a little dog whistle."

Part 5: Ice Water Test

That night Shepard visited a Los Angeles hospital under an assumed name, Victor Poulos, as suggested by the doctor's Greek office nurse.

Louise Shepard sat bedside at the hospital with her husband.

The nurse drew a syringe of ice water from a large ice-filled bucket next to the bed. She stirred the ice noisily to mix the temperature.

"Captain Shepard, this will be your ice water test," said the nurse. She held up the syringe of ice water to the light to check the volume.

"Cucumber Al, I heard. Well," said the nurse, laughing at the absurdity: the first American in space was considered to have ice in his veins. "Let's see if the ice can stay in contact with your eardrum. You've got to stay chilled for thirty seconds."

"This squiggly line here," said the nurse, pointing to a medical device screen, "will show us exactly if our cool customer looks good on an electronystagmography."

She injected the ice water into Shepard's ear. He winced, then grabbed Louise's arm.

"Excuse me," said the nurse, removing his hand from Louise.

"Son-of-a...," said Shepard, suddenly nauseous. "Hey, who's spinning the bed?"

The needle on the electronystagmography began to jump wildly.

"You may feel something like the headache you get when you guzzle a cold drink," said the nurse. She emptied the syringe into Shepard's head.

Shepard tried to keep his voice steady. "Yeah, no kidding." He turned to Louise, "I want this done in absolute secrecy. Talk to the doctor only."

Dr. House entered Shepard's room from around the corner.

"Secrecy huh?" said House. "What name do we use for Captain Shepard?

"Give me any name. Just so long as it's not mine," said Shepard.

Part 6: Extrasensory Perception

Edgar Mitchell's Florida apartment was quiet at night as he spoke on the phone to a drafting engineer, Olof Jonsson, in Chicago. In addition to being an engineer like Mitchell, Jonsson had a deep interest in the powers of the human mind.

Mitchell shuffled a deck of cards having a sequence of symbols written on each one, including a cross, a square, a circle and a star. Mitchell had the soft voice of a late night FM-radio announcer.

"Olof, you're wrong," said Mitchell. "My top card was a star."

Jonsson offered reassurance: "It's only session 10. You've got to concentrate. Send me the thought signals. Whatever's on your card."

"My tally has us 50% right," said Mitchell. "That's double the odds if we were just guessing."

"It's not like telepathy from Florida to Chicago, a thousand miles," said Jonsson, "is going to be anything like from here to the moon."

"We have to synchronize our watches. After liftoff, I'll read the cards, send the thought images to you in Chicago," said Mitchell. "Hell Olof, we're talking about engineering a train of thought to go a quarter million miles."

"Ed, we can drop the idea."

"It's not any skittishness, it's just," said Mitchell.

"Look, I was in the Navy during the war. You can imagine…" Jonsson hesitated. "The ESP soldier."

"I've raised my share of eyebrows. That's not it," said Mitchell.

"What if…" said Jonsson trying to grasp the enormity of leaving the Earth. "This undertaking, this going to the moon and all, it's big. Real big. From Kennedy to now, it's been one question after another, all leading to…" Jonsson paused with this thought. "So Ed, we'll do whatever you think."

Mitchell was silent.

"Hey. Great experiment!" interrupted Jonsson with a sudden insight. "Now I can see very clearly." Jonsson paused with some concentration. "Your star."

Part 7: Recovery

In his Los Angeles' hospital, Shepard wore a large bandage covering the left side of his face and a thin tube draining fluid from his left ear and spinal column. Louise was bed-side as the nurse entered.

"Mr. Victor Poulos, spinal tap," said the nurse reading Shepard's medical chart.

"Poulos? Does that make me Mrs.?" said Louise.

Shepard spoke groggily: "Just don't tell anyone that America's first man in space is flat on his back. With a siphon draining his head."

"How soon will you know?" asked Louise.

"It could take months," said the nurse.

"It's already taken five years. Victor Poulos has a ladder waiting for him," said Louise.

Part 8: Future Test

Early in the morning, just past sunrise, Edgar Mitchell walked the length of the Mission Control parking lot with deep, intense concentration. He read a set of ESP cards with a star, square and circle. He jotted down some numbers on a small scratch pad in his hand. He walked behind a car, whose bumper sticker had become known within the tight-knit Clearlake, Texas community: "APOLLO 14: ONE GIANT LEAP FOR UNEMPLOYMENT."

Part 9: The Cure

Dr. House entered Shepard's examination room with NASA doctors in attendance. Shepard still wore a baseball cap to cover his left side and the small bandage which still marked his surgical incision.

"Come on. Out with it," said Shepard impatiently.

The doctors look cautiously at the floor.

"I've had my share of busted wings. More than any doctor could repair," said Shepard.

"My friend" began House. "You're cured!"

"Don't kid with an old man," said the disbelieving Shepard.

Shepard looked questioningly at the important people in the room with Dr. House, those NASA doctors who could determine whether he is fit to fly.

"Dr. House is correct." The NASA doctor seemed to exude a momentary competitive flash. "Apparently, you are cured."

Shepard steadied himself on the desk and stood straight.

"It's a miracle of some kind." House watched the progress of his important patient. "Full recovery."

"Which means?" said Shepard still unsure of whether to turn his head rapidly enough to verify what the doctors were saying to him.

"It means Al, you are one fully qualified astronaut," said the NASA doctor. "Ready for space flight."

Shepard congratulated Dr. House, snapping his fingers ecstatically by his ear. He turned to Deke Slayton, flight assignment director for astronauts.

"That's it. Damn it, Deke! Nobody's an astronaut," said Shepard. "Not without a flight."

Part 10: The Staircase

The NASA simulators for any flight included mock-up controls and pitch, roll and yaw features that turned a stationary room into an obstacle avoidance chamber. Stuart Roosa, the Command Module pilot for Apollo 14 was talking to Jim Lovell, flight commander on the heroic, but abbreviated Apollo 13.

"The engineers for Fourteen say you'll now have five extra gallons of drinking water," said Lovell. "Fresh off the radiators."

Roosa replied cautiously: "They really know how to get you thinking about water for a one way trip." Roosa paused to consider the reality of his situation and the seat in the simulator. "Best thing is with all this scrambling they must really mean it's true. We're going back to the moon."

"Oh believe me, they're serious," said Lovell. "And once you volunteer to go, what are you going to do? Blame the moon?"

Alan Shepard was working out in the NASA gymnasium at night. He jumped rope, did sit-ups and strained while pulling up on a bar. The gym was empty and he was alone with a somewhat recurrent sound of ear ringing. Sweat covered his NASA shirt. He punched a stop-watch, checked his pulse while holding his breath for fifteen seconds and then forcibly exhaled everything in his lungs.

When he wiped his face with his shirt, a series of tattoo marks revealed the placement diagram for his medical diagnostics and where the NASA biomedical sensors will attach on his chest for flight. He dropped to one knee with exhaustion, and then looked up cautiously at a steep, darkened flight of stairs leading out of the NASA gymnasium. The stairs looked like a canyon to a dizzy man.

The next morning, Shepard was early at the NASA flight surgeon's office.

The NASA surgeon spoke to Deke Slayton, "For any age, Al is a remarkable specimen."

Shepard took Slayton aside and whispered. "Hell, Deke, don't give me that, 'I'll get right on it.' You want it as bad as I do. It'll be the first Mercury 7 astronaut to get to the moon."

"Easier said than…," said Slayton. He paused in thought.

"You remember Mercury, what they used to pack in our astronaut survival gear in case we got lost?" said Shepard.

"Yeah O.K," said Slayton. "Compass. Clean underwear. Sunglasses."

"And a blank check," said Shepard. "US tender. Reserve currency. Good no matter what country."

Shepard and Slayton walked in a huddle down the corridor and continued their conversation outside the flight surgeon's earshot. "And now I'm cashing it!"

"Nobody's lining up to take a backseat," said Slayton. "It's a normal rotation. Everybody will know who's supposed to be next."

Shepard grinned like the cat with a canary, as he began to believe the possibilities. "You tell anybody yet who's got Fourteen?"

"Not yet," said Slayton.

"And we're not going to," said Shepard, shaking hands with Slayton. "So make it official. I'm taking it."

The two astronauts entered the men's room on the side of the corridor. Shepard went to the urinal. Slayton began to wash his face near the sink. One door was shut to one of the bathroom stalls, although neither men could see that behind the door was Stuart Roosa.

"And at the same time you can assign the next guy on your list to 15," said Shepard.

"Damn it Al! Assuming we can get you to the moon, we don't even know if you can climb the ladder," said Slayton.

"You heard the doctor. He likes the specimen," said Shepard.

"Al, imagine it," said Slayton. "You're descending a ladder a quarter million miles from home. The whole world is watching. If you gag in that bubble helmet you're dead."

Shepard was silent.

"You could choke on your own vomit," said Slayton.

"Deke, you're my best friend. But I cannot tell you what it's like. One day you fly around the planet. The next day you've got to stand before a flight of stairs..," said Shepard. "And have those stairs look like a big black hole in the ground."

Slayton splashed cold water on his face. Shepard's eyes were already ice-blue.

"O.K, even without the stairs, my heart's skipping a beat," said Slayton, pausing before a wink to his friend. Slayton's heart murmur was not medically resolved and his flight status remained outside his grasp. "So we'll see what's possible for that first Greek astronaut. Mister Victor Poulos."

"Victor would be grateful. Unofficially," said Shepard trying to mimic a Greek accent.

"But until the smoke clears, talk to nobody," said Slayton.

The two men left. Stuart Roosa stepped out from behind the closed door of one stall and looked into the mirror.

Part 11: Rookies in the Backseat

At a NASA press conference, Shepard made public his crew selection: two rookies named Stuart Roosa and Edgar Mitchell, otherwise known among the astronauts as "The Brain". "We crowned us a crew to the moon. Meet our new assignment, Stuart Roosa, capsule commander, and Edgar Mitchell, Lunar Module pilot," said Shepard.

A NASA technician spoke to a colleague under his breath, "Big Al, he's unbelievable. He picks two rookies to fly backseat to the moon. So no matter what, the elder statesman, Al, he gets his command ride and one very nice and leisurely moonwalk."

"Leisurely? Tell that to 13," said the NASA colleague.

"Anyway in a NASA telephone book, Shepard couldn't have found three more different guys," said the technician. "You know Jekyll and Hyde? In comparison to this crew, they were twins."

Among the gathering of astronauts, hushed accusations flew around that Shepard had picked two rookies to ensure that he would be appointed commander of the mission and therefore he would get to walk on the moon for certain.

Stafford said, "Rookies in the backseat."

"Wrong," said Shepard unapologetically. Shepard put his arms jovially around his shell-shocked crewmates. "I picked these guys for one reason only. They're tops. They can do the job. And I want them with me. Do you guys read me?" said Shepard.

Cautiously the gathered astronauts deferred to the senior Shepard begrudgingly.

Stafford said, "Five by five, Al."

A similar scene played out half way around the Earth. In a Soviet Mission Control Room, the NASA counterpart, Alexander Komski, was leading a robotic mission to the moon. He too was conducting a press conference for western reporters.

"The Union of Soviet Socialist Republic is pleased to announce the fifth successful month of operation for our robotic moon explorer, *Lunokhod I*," said Komski. He began to point at a series of engineering drawings. "*Lunokhod I* weighs nearly a ton."

A news reporter wrote quickly, while speaking under his breath, "Damn monster truck on the moon."

"It has eight-wheel maneuverability and will navigate near the range known as *Fra Mauro*. It is thought to contain some of the oldest geological records," said Komski. "Both on the moon and in the solar system."

"Excuse me," interrupted a second news reporter. "So in the international weight-lifting competition, are you putting *Lunokhod* as the Soviet's newest contestant?"

"From the Soviet side, there is no contest," answered Komski confidently

The news reporter continued, "Can you comment then on the Apollo 14 mission now headed for a similar landing site?"

"Obviously, we will exchange some soil samples with them after they return," Komski said. "And I might remind you, if we can provide any rescue services, as we offered on your recently failed 13 mission, our scientific and engineering experts will be available in whatever capacity."

Part 12: Engaged

The astronauts, Ed Mitchell and Alan Shepard, were training in the Lunar Module simulator. A huge landscape of the moon appeared in the window. The camera was inside, such that the simulation was indistinguishable from the real landing--the bright lunar surface appeared to move by the windows quickly. Its pockmarked surface was bright and dotted with craters and rough terrain.

"Lunar descent now engaged. Landing radar is up. LM 14,000 feet," said Mitchell referring to the Lunar Module or LM which he was to pilot.

"Radar up. PGA, blue to blue. Do we have a purge valve?" said Shepard.

"Lock-locked and verify Low. Visibility good," said Mitchell.

"Got Low. Computer. Do you see where the apple is?" said Shepard.

Mitchell looked to the window and scanned the lunar terrain.

"Got PGA diverter valves vertical." Mitchell did a double-take on the window scenery flying by his field of view. "Son-of-a-bitch!"

Shepard looked momentarily at his checklist, as if 'son-of-a-bitch' was an unanticipated flight command. Shepard squinted "What the…?"

In the foreground, outside the window was a huge, bug-eyed space monster on the surface of the moon. The shadow of the creature enlarged to huge dimensions.

Outside the simulator, a technician, John Von Bochel, had stuck a dead horsefly on a pin in front of the simulation camera. The camera projected the moon's image for realism and now outlined the shadow of the bug-eyed horsefly onto the lunar panorama.

Mitchell dropped his hands from the controls and began to breathe again. Shepard ran an open palm across his scalp.

"Anybody got a gun?" said Shepard.

A cadre of technicians laughed in the background.

Slayton entered the simulation room. Behind Slayton follows Bob Hope, who carried a golf club as his walking cane, to the simulators and the rocket mockups. Shepard, Mitchell and Roosa were doing flight simulations in and around the Lunar Module. Slayton knocked on the simulator door.

"So, Mr. Hope," said Slayton pointing to the Lunar Module, "this is what will land us on the moon."

Hope held a golf ball up to the light, to make it look like the moon. "Small target."

Slayton made his counterpoint, "Bigger fairway."

Slayton knocked again on the door. From inside the Lunar Module simulator, Mitchell turned to Shepard unbelieving about the repeated number of interruptions.

"I'll keep running it in here," said Mitchell.

Shepard went out to greet Bob Hope briskly.

"Deke, we got less than a hummingbird's heartbeat before we're going to the moon," said Shepard. He pauses. "Oh, hello, Mr. Hope. We met in...uh, when was it? 1961?"

Bob Hope lifted a golf club in greeting. "May, 61. In your New York parade. First American in space. Enough confetti falling that day to plaster New York in summer snow. Never seen JFK so excited." Hope paused with the timing of a professional master of ceremony. "Not without a blonde in sight."

"This is going to be a full-up mission," said Shepard.

"You guys got that exploding oxygen covered from 13 yet?" asked Hope.

Slayton interrupted the train of thought, preparing to escort the visitors on the rest of a tour. "Uh, Mr. Hope..," said Slayton.

Mitchell said from inside the Lunar Module simulator. "Hey, Al, the LEM is showing a low thrust ratio above the Sea of Rains."

Shepard winked at Hope and his golf club. "Sorry, Mr. Hope, my tee-time's being called," said Shepard. "We're driving this bug to the moon."

Part 13: The Pick

The NASA press corps gathered to question the Apollo 14 crew.

A reporter directs a question to Edgar Mitchell, "Ed, a lot of American are wondering, as a scientist-astronaut, what do you hope to learn from going to the moon?"

Mitchell said, "People think it is all about space travel. But it's time travel." Mitchell paused. "These hands will gather rocks and soil that has not been disturbed since the beginning of the Earth." Edgar Mitchell could switch from engineering to geology to the Bible in mid-sentence. "When Eden was, these rocks were."

"Most people don't grasp that the moon controls things like the ocean tides, even female fertility," said Roosa. "I got four mighty grateful kids thanking the moon."

A second reporter directed his question to the commander, "Captain Shepard, as the first American in space, do you have any comment on the Russians calling your first flight," he flips in his notebook for the exact phrase. "I think the quote was 'a little flea hop'?"

"Honestly, what did I think ten years ago? Sure I thought I'd be going to the moon," said Shepard.

The news reporter continued, "But if I can follow up, Jim Lovell on Apollo 13 said that with all their problems, Thirteen would be the last flight to the moon for a long time."

Shepard smiled his test pilot's grin, "There's no greater thrill in life than to be shot at and missed."

"But for some reason, God forbid," said the reporter, "if this flight does not meet all its scientific objectives?"

"Let me tell you something about public and private opinion," said Shepard. "When I was the first American in space, the Republicans approached me to run for President. Can you believe it? 16 minutes on a candlestick and I'm qualified to run the country. That's public opinion," said Shepard. "But then for 8 years, I lost my flight status. Something as tough as falling down. A buildup of less than a teaspoon of fluid in my ear and I was told I'd never fly again. A death sentence to a pilot. It grounded me to flying a desk. No public opinion there, just a private truth So we're going to the moon. This country decided to plant a flag at *Fra Mauro*, that's in the Sea of Rains, and we'll be walking in a meteor's footprint the size of Houston. That crater, practically speaking, predates dinosaur days. So if 13 can't do it, then 14 will. And if 14 can't do it, then 15 will."

Another news reporter changed tack, "So Captain Shepard, how does it feel to be the oldest astronaut to fly in space?"

"To be honest, my secret: wake up next to a young wife," said Shepard.

"With your ear problem and losing flight status, people are calling you one of the comeback heroes of all time."

"If I can fit into it, I'll fly it," said Shepard. "But I suppose if we don't make it back to earth, somebody will say the poor son of a bitch wasn't ready. But I am ready."

"Commander Mitchell is it true that you have had some psychic experiences?" said another reporter.

The question created a certain tension in the room. Shepard frowns as if the question is another joke. But Mitchell answers matter-of-factly: "By psychic, you may mean the recognition of sensations by extrasensory perception?"

"ESP?" mimicked Slayton. "Well, uh..."

"These perceptions," said Mitchell. "They anecdotally involve bright light. Subjective sound. Hearing hallucination." Mitchell seemed to have the symptomatic checklist memorized in his head. 'And in some rare cases, thought communication across long distances."

Slayton got into the swing of the conversation in his characteristic way of deflecting uncomfortable moments with humor, "You guys know: brainwaves."

"Unfortunately its prediction success has not been verified statistically," said Mitchell. "So that means we don't understand it enough at the moment." He looked at his surprised colleagues and reporters who remain skeptical. "So in my scientific opinion, we don't understand it well enough yet to live by it."

The Press Room fell silent.

"What do you mean exactly by 'yet'?" asked a reporter sensing that a news headline was about to be written. "Assuming 'it' exists?"

Shepard quickly covered any awkwardness. "Heck, I've heard ringing noises ever since my Mercury days," said Shepard.

Jim Lovell, commander of the failed Apollo 13 mission, entered the backdoor and looked on.

"But people are still asking: is Apollo 14 another Moondoggle?" continued another reporter borrowing the newest slang phrase to punctuate the public description of a boondoggle.

"Excuse me," said Lovell interrupted from the back of the room.

The NASA press corps turned to the back of the room and flash random pictures of Lovell.

"I'll testify. This flying to the moon, it's no bus ride," said Lovell.

"Further questions?" said Slayton.

"With each flight, fewer Americans are watching …" said another reporter.

"Watching? Heck, I'll be watching all alone on the back side of the moon," said Roosa. "No complaint here. Just tapping my fingers to Hank Williams' songs." He continued with an afterthought, "Anyways, how many people saw Columbus land?" Roosa paused. "Four Indians and a squirrel."

A NASA Public Affairs Officer steps to the podium. "Thank you, gentlemen."

The NASA Administrator summarized the press conference. "Not to put any added pressure on these fine Americans, but let me just add one thing. If Apollo 14 doesn't go well, we may not have a future at all. I feel, strongly that Apollo 14 has got to be a perfect mission."

Roosa and Mitchell left the stage after the press conference had dismissed and spoke quietly.

"Perfect," said Roosa, shaking his head. "No such thing. You ever seen a perfect mission in your life?"

"Only once," said Mitchell.

"I don't believe anybody can," said Roosa.

"In the mind, your mission can always be exactly, precisely...perfect," said Mitchell.

"Hell, I was trained to fly bombers deep into Russian airspace, then eject from my plane before the bomb was delivered," said Roosa. "Survive off of roots and berries while hiking out of whatever fragments of Siberia my bomber left behind. That couldn't be done perfectly." Roosa had trained as a pilot for spy mission deep in Soviet air space. "And the moon is not Siberia."

"This moon trip, it can be about something big, about evolving our humanity," said Mitchell.

"Hell Ed, stow it," said Roosa.

"No, I'm telling you. I met a guy in Chicago," said Mitchell. "You see this guy's a skipper in World War II. Jonsson's his name. His ship, all his buddies, got trapped in a minefield with no way out. But this is it: purely with thought, the guy guides his commander and gets them all past where each of twenty underwater mines are waiting."

"Yeah?" said Roosa skeptically.

"It can't be faked. 20 undetected minefields and his whole crew gets safely out of that briar patch. He simply...well, divines where every explosive charge is," said Mitchell.

"Too much, buddy," said Roosa. "You just divine us to the moon. In about 15 days, I'll listen to anything you got for me." Roosa had lost the edge in his voice. "While sitting in your Houston hot-tub."

Mitchell got ready to leave. "I already talked to Shepard" said Mitchell. The mention of the hot-tub made him consider his house. "My wife's fed up with the long hours."

"The moon will be easier than your old cowboy days. Like New Mexico. Just rounding up cattle at the end of the day," said Roosa. "It's time to get on that horse again. To the moon."

Part 14: The Grandfather

In the San Gabriel Mountains, California, it was a hot afternoon.

A geology class was being run by Caltech Professor of Geology, Lee Silver. Dressed in desert gear, the three Apollo 14 astronauts arrived. Shepard, Mitchell and Roosa drove over the rough and rugged terrain along with Silver. The professor was the grandfather of geology and could interest anyone in the study of rocks.

"Men, you're about to enter the lunatic asylum!" said Silver. He was driving a truck very quickly through the desert.

"Every dust particle, every lunar boulder, it all comes through these hands," continued Silver.

Silver took a turn quickly, almost spilling the truck and the astronauts on the desert floor.

"Flying, that's your asylum, "said Silver. "Caltech's mine. And the lunatics are us scientists wailing endlessly at that moon."

The truck came to an abrupt stop and a cloud of dust blew forward in the tracks.

"So you turn us pilots into geologists," said Mitchell.

"At the picnic you see," said Silver, "we're the anteaters."

"I figure a real geologist has got a lifetime to study the dust we bring back," said Shepard.

Silver was perturbed at anyone dismissing his life's work. "But what I study for the rest of my life, that's what you pilots are going to have to collect from the moon." Silver paused. "So from where I sit, when this old mountain is through with you..." Silver points to the summit ahead. "...you flyboys are going to be master gem cutters."

"Yeah right." Shepard said sarcastically. "A diamond from a cow turd."

Silver ignored Shepard's acerbic comment and remained ever enthusiastic: "You pilots may think Apollo 13 ran into trouble. Most interesting mission ever to me."

"You kidding?" said Roosa.

"Hell, no!" said Silver. "When wreckage from ol' Jim Lovell's ship crash-landed on the moon, set off a damn symphony of moonquakes. Sent our seismographs ringing like a church bell."

"Huh?" said Shepard.

"I got half a dozen magnetic tapes off that albatross. 13," said Silver. "You should be so lucky to fail."

"Whoosh," said Mitchell.

"Hell, you guys may want to fly against the Russians?" said Silver. "But for us at Caltech, we're just flying against the pinheads at the Smithsonian. Only edge we got, those damn museum people don't catch the moonquakes." Silver smiled to himself. "That's our card game."

Silver cracked his knuckles and spit out the window. "So let everybody else eat our dust. Literally."

"We'll work it, Professor," said Mitchell.

Silver stopped the car. "Good, then you've got 10 minutes. Find me the suite."

"The suite?" said Mitchell.

"A dozen hand-sized rock samples. From the garden variety granite to the exotic gem," said Silver.

"Ten minutes? You're kidding," said Roosa.

"Out there?" said Shepard.

"It won't be any easier on the moon," said Silver. "You pilots are trained to observe, so just look."

The three astronauts trudged off into the Sonora Desert, while Silver popped open a can of soda and lowered his hat to shade his eyes from the intense sunlight.

Shepard stepped out of the truck and muttered under his breath. "Damn Easter egg hunt."

"Captain Shepard!" Silver offered Shepard a cigarette which in the heat, Shepard waved off.

"Your NASA papers say you're some kind of genius." said Silver. "IQ of 145! Even sometimes like to go boating on one ski." Silver paused. "But get wise, these moonwalks, they'll be barefoot, not a ski in sight and without a tow rope."

Silver raised his hat brim to reveal shiningly clear blue eyes

"You remember Apollo 11?" asked Silver. "Buzz Aldrin, the second moon man? He had the whole world listening when he said he found some interesting rocks on the moon. Said they looked like biotite."

"Nice piece of geology, if you ask me," said Shepard.

"Gentlemen, biotite is a variety of mica. It forms only in the presence of water. Water means life."

"So?" said Mitchell.

"So do you want Apollo 14, America's hope for a moon program." said Silver. "To announce to three-quarters of the planet-including your mother, your third-grade teacher and the President-that you, Al Shepard and Ed Mitchell, the nation's finest, have succeeded in finding biotite?"

Silver let the thought sink into the three astronauts. "And therefore life on the moon?"

Shepard and Mitchell gave each other cautious looks.

"Nine minutes, gentlemen," said the professor.

"Buzz didn't say it was biotite. He said it looked like biotite," said Shepard defensively to Mitchell.

The two began to search for rock varieties. Shepard was winded and began to sweat profusely.

Lowering his hat over his eyes, Silver said, "Biotite, ha!" He laughed to himself. "Damn aliens on the moon. Report that one to the Pentagon."

Part 15: The Quarantine

Slayton walked the cat walk the night before launch of Apollo 14. He walked the tower and kicked the tires. A huge moon silhouette could be seen in the background.

A NASA doctor, Dr. Cleman stood before the astronaut crew, discussing the plan to quarantine the Apollo 14 crew for 21 days.

"This is it. We're closing quarters, gentlemen," said Dr. Cleman.

"Do you know how long a 21 day quarantine is, doctor?" said Roosa.

"Exactly three weeks," said Dr. Cleman.

"I don't think you understand, I'm going to be alone, completely alone on the dark side of the moon, with no one. And you're saying I have to...I mean, I've got 4 kids," said Roosa.

"And you plan to see those kids deliver grandkids," said Dr. Cleman. "Gentlemen we have two problems. First, Apollo 13 lost a crewmember to measles exposure less than two days before launch. That will not happen on 14." Cleman paused. "Secondly, after skating by a direct lightning strike at launch, Apollo 12 brought back some kind of moon bug."

"I don't believe it," said Shepard.

"Not exactly a virus or some kind of *E. coli.*," said Dr. Cleman. "But it was a genuine sample of a strain called *Streptococcus mitis.*"

"Life on another planet?" asked Mitchell. "We've been through this lesson."

"Not precisely extraterrestrial," said Dr. Cleman." *Streptococcus* was an earth grown bacteria. It hitchhiked a ride on the Surveyor spacecraft. And after that spacecraft robotically landed on the moon, Apollo 12 brought back a chunk of that ship, and believe me, damned if we didn't find life on the moon."

"Slugging away in minus 200 degree weather?" said Mitchell still unbelieving.

"It's true, we've unknowingly," said Dr. Cleman. "Seeded another planet."

"There's no water of course," said Dr. Cleman. "At least, not that we know of and the damn critter was freeze-dried on the Surveyor's antenna. Almost glued there but nevertheless, the strain did survive for several years." Cleman paused. "On the moon."

"So for that, one bug, we're off limits from earthlings for 21 days," said Roosa.

"Correct. No discussion. The doors will close at midnight," said Dr. Cleman. "I suggest you not be left outside with only one slipper."

"Louise?" said Shepard asking about conjugal visits.

"Wives only as visitors," said Dr. Cleman. "But there will be glass."

That night before the launch, the three astronauts stood in the lobby of the NASA Quarantine room separated by glass. The astronauts talked with their wives and children. Shepard and his wife Louise both pressed their lips to the glass, and then Shepard stepped back from the quarantine partition.

Louise eyed the Apollo 14 patch on his flight suit. It showed the 14 crew carrying the whole astronaut office with them to the moon.

"Nice patch. Take care of it, will you," said Louise.

"If...if for some reason... I can't return it," said Shepard. He pushed his hand to the glass. "You and I, we..." Shepard fumbled for the words. "Ever since the *Letterman's Ball* at Annapolis."

"You asked me. One problem, you didn't have a varsity letter," said Louise.

"Nay-sayers," said Shepard defiantly.

"You got your letter... I never thought about saying anything but yes to the letterman," said Louise.

"Now we've got four letters: NASA," said Shepard.

"You asking me for another date?" asked Louise.

"I won't be making my usual phone call tomorrow night," said Shepard.

Louise looked nervously at the ground.

"I'll be leaving town," said Shepard.

Roosa's wife, Joan, touched the glass separation. She spoke to her husband, nicknamed "Smokey" for his days as a forest fire fighter.

"You bears don't start any forest fires up there," said Joan recalling the explosion on Apollo 13 and fire on Apollo 1." Joan winked a smile. "We'll be waiting, Smokey."

Chapter 2

Part 1: The Clock Starts

NASA Flight Controller, Gene Kranz, waited in Mission Control for the delivery of his trademark white vest made by his wife. Capsule Communications Office, CAPCOM, entered the row of computers and big screens as he carried a double box of cigars.

"I hate launching in the afternoon. Where's the moon?" said Kranz. "You can't even see what you're aiming at."

Kranz put his hand on the shoulder of a very young flight controller, the man in charge of electrical circuitry, otherwise known as EECOM.

"You shaving yet son? This must be your first flight," said Kranz.

"Yes, sir," said EECOM.

"Well, belly up to the console, son. Your predecessor, his shadow's still warm in that chair," said Kranz.

EECOM shifted uncomfortably. The average age of a NASA worker was still under thirty.

"He lost his seat in this room, why?" said Kranz.

"Because he ran his car into a brick wall, as I understand," said EECOM.

"Anything else?" said Kranz.

EECOM was unsure what the right answer was. And when Gene Kranz asked a question, anyone in Mission Control had to know the right answer or propose an alternative before the next question was posed.

"He survived without a scratch. Lucky fellow. He wasn't even reckless. I asked him to step down for one reason," said Kranz. "He couldn't give me a clear explanation of why the accident happened.

"I feel great lucidity coming on, Mr. Kranz," said EECOM.

"There's only one failure in this room," said Kranz announcing the edict to the line of young engineers filing into Mission Control. "The inability to duplicate the last failure."

Kranz paused. "That means we have no earthly idea what's going wrong."

Kranz dropped a heavy flight manual on his table.

"So just remember this, when the papers' weight equals the rocket's weight, it's got to be time to launch," said Kranz.

Kranz put on his characteristic vest to applause. He put his hand on the congratulatory cigar boxes. Underneath the glass on his desk, he slid an eight-by-ten photograph of the blown-out side of the Apollo 13 Command Module--the blown out oxygen tank, the cryogenic tubes and severe burn marks..

"Gentlemen, start your clocks," said Kranz. "This is Apollo 14. We're going to the moon."

Part 2: Launch Day

On the launch day at Kennedy Space Center, the astronauts filed down a walkway which was empty of observers and press.

"It's like a ghost town," said Shepard.

"Had to clear away the onlookers," said Slayton. "No virus exposure."

Slayton pointed at clouds and looked to the sky for lightning. "Those clouds might move into the launch area," said Slayton.

"We're clear," said Roosa confidently.

"Have a good trip," said Slayton. He flashes his farewell, with thumbs up. "And watch your ass."

Slayton says to himself. "The rest of the world will be."

As the astronauts talk with engineers, they are strapped into their seats for launch. Changes since Apollo 13 checkout were reviewed by Shepard with a list in his hand.

"We've given you 160,000 changes," said a technician.

"Just give me one change," interrupted Shepard. "A moon landing."

"Got it," said the technician with a slight German accent.

"Prime the oxygen tanks," said Shepard.

"Two of them before," said the technician. "Now we got three."

Shepard tried to burn off stress in conversation. "Good going. Two failed before," said Shepard. "So a slide-rule gives us what? A third chance not to cock it up."

"The tanks are all isolated. A quick disconnect. You don't even have to hold your breath," said the technician.

"Just relax and look up," said Roosa exhaling.

The technician pointed a flashlight towards the interior of the Command Module. Light flickered briefly and then shone brightly.

"To boot, you got a spare battery. High current," said the technician. "Four hundred extra amps."

"Don't plan to leave the headlights on, thank you," said Mitchell.

"Button us up. More of everything, we've got. Only we're getting fewer chances at the moon," said Shepard. "Let's go."

"Are we set?" said the technician in a more heavily accented German. *"Iss .. good locomotive, but vhere are the tracks?"*

"Apollo 1 through 13," said Shepard pointing up. "Right there is your track."

The technician strapped in Roosa very tightly, the last one in.

Part 3: Magnus Maximilian

"T-minus nine minutes." At the Kennedy Space Center, a voice on a loudspeaker was counting down to the ignition of Apollo 14. "Green all the way."

"Nine years of waiting " Mitchell turned to Shepard in the Command Module. "And hello!"

"This is a snap," said Shepard. "In '61, the flight doctors had me wired. They even knew I had one toe nail-fourth toe-coming off in flight. From an engineer stepping on my foot." He laughed to himself. "Told the whole world, old Al Shepard's rectal temperature just went up a degree."

Shepard paused to relieve his historic flight. "And of course, adrenaline. That's the booster fuel. They said for fifteen minutes, my adrenal glands opened like a fire hose. 250 percent above normal. It was pure octane."

Shepard laughed louder to himself. "And of course, I peed in my suit. Filled my shoes. Actually up to my ankles."

"Give that monkey a banana," said Roosa laughing.

"Damn near shorted out the suit," said Shepard.

Shepard checked his seat restraint. He was suddenly all business now, "Okay buster. You volunteered for this thing."

Shepard put his hands firmly on the arm-rest, slowly repeating to himself. "X-squared-plus-Y-squared..," said Shepard.

Mitchell and Roosa looked on in disbelief.

"Uh, it's the quadratic formula..," said Shepard. "When you're used to flying alone then doing algebra in your head. It keeps your focus."

"You're the boss, Al," said Roosa scratching his head.

The voice on the loudspeaker continued to count down towards launch. "T-minus eight minutes."

"Let's go to the moon," said Roosa.

"And we're holding. That's a hold," interrupted the voice over a loudspeaker.

"Holding," said Shepard. "Repeat, did you say we're keeping the count?"

"Looks like we'll hold for a while for the storm to pass."

"Christ, not again," said Shepard.

"Hey the way it looks, this storm is going to go right over us and out to sea. So make yourselves at home," said Deke Slayton, head of the NASA Astronaut Office.

"Yeah right," said Shepard quickly and sharply. "Tell you what, Deke, let's crank the handle."

German rocket scientist, Wernher Magnus Maximilian Freiherr von Braun, or Dr. von Braun, stood in the background of Kennedy Space Center launch control room and pointed to the sky. The east coast of Florida was being lashed by pounding rain. Over a loudspeaker, a launch technician announced, "Count is T-minus eight minutes and holding."

"Got the hold," repeated a second launch technician.

The mission clock did not move from the printed numbers of the countdown. The launch technician put down his earphone and remote microphone and said to his colleague, "Shepard, he's got some kind of antifreeze in the veins; you know we ran into the same launch delays then too, back in 1961. So what does Cucumber Al say from inside the belly of that rocket?"

The second launch technician shook his head unknowingly.

"I'm cooler than you are. Why don't you fix your little problem and light this candle?" said the first launch technician. "With 100 million Americans listening in, the guy comes live over the radio. He says to the world, light this candle." He pauses remembering the scene from nearly a decade ago. "Now that's a guy who's pure band-saw steel."

Part 4: The Storm

Edgar Mitchell scanned some gauges inside the Command Module atop the immense Saturn V rocket. "Uh ...Kennedy, while we're holding a wet finger to track the storm, how's our, uh... water coolant holding," said Mitchell.

"Look there. Dumping something like 80 swimming pools on this fire stick every minute," said Shepard.

"Clouds above. Clouds below. Get this rocket off the stand," said Mitchell.

Roosa looked out the window as the storm departed. "Uh...Kennedy, it's not like I've got a rear-view mirror or anything ... But either that storm's moved out to sea... or we've already launched," said Roosa.

"I don't know," said Shepard. "In flight school, we said engineering is like diarrhea: it just keeps dribbling on forever."

"T-minus eight minutes and counting," said the loudspeaker announcing the launch countdown had begun again.

"Let's stoke the furnace," said Shepard.

The launch technician continued coolly, "Initiate firing command."

On the Rocket Test Stand the blue sky broke through cloud cover. The rocket began its preparatory sequencing as oxygen vented and ignition steps began.

Wernher von Braun looks on with anticipation from the Rocket Firing Blockhouse. Von Braun began to pray almost silently to himself. "So... Our Father who art in heaven."

From her seat at Kennedy Visitor's Stand, Edgar Mitchell's wife, Louise, concentrated on the rocket in the distance.

Mrs. Roosa said to her husband, "Come on, Stu. Go!" Her eyes well up with tears. Her four children put their arms, octopus-like, around her skirt.

In the Command Module a quarter mile away, Shepard read off his checklist, "Initiate sequence start."

A voice of the launch technician continued with a metronomic efficiency, "T-minus 8 seconds and counting ...7 ... 6... 5.....4 ....3.... 2.... 1. All engines running.. Zero. Ignition!"

A low roar in the distance made pelicans scatter. Florida swamp water began to form waves. Alligators went underneath the surface. A tremendous shock wave hit the crowd as their pant legs began to ripple from the vibration. The rocket appeared from a deep rising cloud of smoke and acetylene fuel.

The Apollo 14 crew shook violently in the rocket cabin.

"Stand by for the train wreck," said Shepard.

Shepard announced the separation of the bottom rocket stage as it climbed skyward. The astronauts jerked forward in their seats like a speeding locomotive hitting a wall.

Edgar Mitchell settled back and undid his seat restraint as they entered low earth orbit. A loose screw floated by in the zero-gravity cabin.

Shepard snatched the screw out of the air. "Always one screw loose," he said.

Mitchell took his first look out the window as they passed over the Himalayas. "Will you look at that? There's smoke coming out of those chimneys," he said.

"It's Tibet. The Himalayas are about all the geography you'll see from the moon," said Roosa.

"Houston, I think we smashed some bugs on the way up." Shepard looked with bewilderment. "I've got blood on the window," he said.

"Just make sure it's outside," said Kranz.

"Fourteen, you're 'Go' for trans-lunar injection," said the Capsule Communication Officer, or CAPCOM.

Mitchell read an altitude gauge. "We're climbing like a squirrel shimmies a tree trunk," he said.

Stu Roosa was exuberant. "Punch my ticket... for the moon," said Roosa.

"Valet parking. Fully validated," said Shepard.

Roosa enjoyed the sudden free-floating weightlessness for the first time. "Hell, Al, at this moment I couldn't even tell you where to put the car keys," said Roosa.

Mitchell took a moment to figure numbers on a scratch pad. "Even Apollo 13 made it this far," said Mitchell.

"Houston, beginning barbecue roll," said Roosa.

The Command Module began to rotate slowly to relieve the excess heat buildup from the sunward side of the ship.

Houston CAPCOM breaks into conversation as the earth looms smaller below the three men.

"We copy. You have a Go, Kitty Hawk," said CAPCOM.

"Let's do it," said Shepard quietly.

"Houston, we're in a position to proceed with docking," said Roosa.

"That's a go," said Kranz.

Command thrusters fired with the touch of pilot Roosa and the Command Module named Kitty Hawk steered into position to dock with the Lunar Module named Antares.

In the field of view out the side window, Roosa eyed his target: a conical shaped collar in which he was tasked with putting the docking probe prior to getting the okay for going to the moon.

"Alright Stu, you hold all the records. Least fuel ever used in a docking," said Shepard. "Show them how 14 meets the road."

"This ship is gonna drink like a camel," said Roosa. "Houston, we have visual contact. The Lunar Module is in our sights."

Edgar Mitchell said to himself, "Pilot's prayer: My God, let me not be ashamed."

Below the module, the beautiful blue ocean and white cloud tops glittered in the reflected sunlight.

"Fuel gauge shows we're green. We're preparing to unstack this layer cake," said Shepard.

"We show docking target is closing," said CAPCOM.

Roosa fired his hand-held thruster to ease the two ships together. The docking maneuver looked picture-book perfect.

"Hot damn! You hit it dead center, Stu," said Mitchell.

"You tell me, who can use less fuel to thread that needle," said Shepard. He paused. "Houston..., uh ...hey, wait a minute." Shepard was bewildered by the readings in front of him. "We have contact but our panel is showing no capture. Can you confirm?"

"We copy. Panel shows no capture," said CAPCOM.

"She's not taking," said Roosa.

"Can you repeat that, Kitty Hawk?" said Kranz.

A long, tense silence fell on the control room as the men leaned back in their chairs, lit cigarettes and ran tired fingers through their hair.

"Damn. Thirteen," said CAPCOM with his microphone turned off. He flipped a switch and spoke again to the crew aboard Apollo 14. "Roger, Kitty Hawk. We show no capture."

On the panels ahead, no green light or sound was shown.

"We just lost the fuel record," said Roosa frustrated from the Command Module.

"The record," said Mitchell, "is going for the moon."

"Houston, the arrow's on the mark. We should've parked this bus," said Roosa.

"She's been simulated a thousand times," said Kranz.

"But this is game day, boss," said CAPCOM.

The right leg of CAPCOM began moving in the classic sewing-machine-style, the bounce of a nervous, waiting man.

"Stu, what's she reading? Copy," said CAPCOM.

"Houston, we've failed to secure a dock," said Roosa.

"Damn," said Slayton. "Fell off that horse again."

A long, pregnant pause hung in the room as the Mission Control team settled in for a long flight.

"Roger, Kitty Hawk," said CAPCOM, suddenly breaking the silence with a firm command. "We'll work the problem."

Kranz spoke again off the microphone "Tell them...even if it doesn't fit...if it only touches ...then let's nail it down."

"Kitty Hawk, you've got a 'Go' for another attempt," said CAPCOM.

Mitchell turned slowly to look out the window at the shrinking earth below.

"Al, what do you think," said Roosa as he pulled off the finger-roll joystick used for guidance.

Shepard asserted his instinctive command, "The position was perfect. So were you." He turned to the window. "Let's go again."

Part 5: Going in Again

Roosa nodded hesitantly. "Houston, we're going in one more time."

Radio crackle and hiss separated the silence in Houston. Shepard and Mitchell watched helplessly as the Command Module was maneuvered into position again. Houston controllers stood from their computers and watched intently on the trajectory board for the rapid movement of the two separated spacecraft.

Far out in space, the thrusters turned the ship towards its docking partner which was required if successful landing on the moon is planned. Mitchell turned to Shepard quietly.

Mitchell said under his breath. "X-squared-plus-Y-squared."

Mitchell tried to distract himself by scribbling some numbers on a pad in front of him. Finally he spoke to Shepard confidentially. "Hard to get to the moon without that free taxi ride waiting for us out there."

"You're the Brain. You figure it out," said Shepard sharply as he points at the list of numbers in Mitchell's hand.

"You want to think us there?" said Roosa mocking of Mitchell's interest in the paranormal.

Shepard said: "Alright." He took a full breath as if preparing himself. "Before they take the moon from me, I'm opening that hatch right there…" He pointed to a heavily bolted door which separated the astronauts from the frighteningly hostile environment of space. "...and I'm crawling across that black canyon between me and that Lunar Module ..." He gestured a spacewalk with his fingers. "...and I'm going to mate this male probe...with that female cone." He gestured about the receptive docking gear. "...and we're making some moon babies."

The Command Module drifted closer to the Lunar Module's docking apparatus. Probe and cone met perfectly and fit together tightly.

Roosa fingered the two ships closer. "Contact."

Inside Mission Control, the young technicians stood motionless.

"Roger, we're waiting for confirmation," said Kranz.

"Negative. Houston," said Roosa. His voice was flat and frustrated. "We do not have a dock."

Silence filled the control room punctuated only by the shaking of nervous heads.

"Copy, Fourteen. No dock," said Kranz.

"We're going to pull back and give this some thought," said Roosa.

"Roger, Kitty Hawk. We'll be doing the same down here," said Kranz.

"Can you confirm our fuel status?" asked Roosa.

"Roger, we'll be on tight watch," said Kranz.

The silence was deafening. After several seconds of silence, a group of backup teams announces suddenly.

57

Slayton took his commanding position over the success of this flight. "Okay, you heard them. Let's scramble a solution. Where the hell is the probe and drogue?"

"Standby teams. Get me the experts on this system. Where's the assembly?" said Kranz shaking head impatiently. "We've always had a docking probe and drogue available."

"No docking, no mission," said CAPCOM to himself, looking helplessly at the control panel.

A hovering Public Affairs Officer said, "You know if anything goes wrong this time, you'll hear the hounds baying at the moon." He paused: "Literally."

"We're going back to the moon," said Kranz affirmatively. "I don't care if we have to fasten a rope out of those guy's urine tubes. I'm telling Al to lasso her."

As the Command Module orbited the Earth, the crew on board grew impatient for a decision.

"Houston, we're waiting for recommendations," said Roosa.

"Roger, Kitty Hawk. Standby," said Kranz.

"Standby?" said Shepard suddenly hitting his armrest.

"We've got to fix this thing," said Roosa.

"Clear as day," said Mitchell flipping pages in a flight control guidebook. "Given the failure of the first docking maneuver."

"No second failure," said Shepard pointing at the closed hatch.

"We need a hitchhike back from the moon, no doubt about it," said Mitchell.

"Houston, it's just a lock and key mechanism. Nothing tricky up here," said Shepard changing the subject impatiently from Mitchell's doubts.

"Roger, Kitty Hawk," said Kranz. "Our boys are working the issue."

A huddle of engineers was repeatedly twisting the probe and cone for a possible locking failure.

"The probes just locks. That's it. No electrical energy. No hydraulics. No pneumatic drive. What can go wrong," said Shepard.

"Tell that to 13," said Roosa distractedly preparing for re-dock.

The day had turned to night in the windowless Houston Mission Control.

"The designers are working the problem, Kitty Hawk. Over," said CAPCOM. He continued off-mike. "Works like a damn skeleton key down here."

A Houston engineer ran his finger along the lip of the cone for rough spot. "Listen, if a piece of debris or even a speck of dirt is lodged in this mechanism."

He pointed to the drogue. "Then it will prevent the latches from depressing.

A group of engineers huddle closer. "That prevents, as well, their snapping it into place," said Slayton.

A second engineer said, "Anybody got a clean handkerchief they can spare?"

"Hell, if debris is the problem, then Stu should keep at it," said the first engineer. "Coming back again could dislodge whatever is blocking the hard dock."

A third engineer concurred, "Yep, he's right. Tell them to stay with it."

"Kitty Hawk," said CAPCOM. "The immediate plan is to maneuver to re-dock. If there's debris blocking the latch, then we're going to keep trying."

"Houston, Roger that," said Roosa. "We'll try a third time."

The third engineer said to his colleagues, "How many times will she take 'no' for an answer?"

Roosa re-maneuvers for docking, and then fails again to see a green light.

"Houston, no joy," said Roosa.

The docking failed again. Shepard waved for a retry, while checking the fuel reserves.

"Can you confirm the fuel is still green line?" asked Shepard.

"Roger, Kitty Hawk. We show reserves adequate," said Kranz.

"From my chair, I give them two-make that one more try-before we hit critical," said an engineer to Kranz.

A third engineer said in disbelief, "Empty?"

CAPCOM ignored the exchange in the interest of not disrupting crew activity, "Kitty Hawk, we show your fuel nominal for another attempt."

The fifth docking attempt failed to lock in again.

"No joy," said Roosa.

"Houston, we confirm. No hard dock," said Shepard.

"The latches aren't clicking," said Kranz.

"One more try. This is the keeper," said the second engineer speaking off-mike from the crew.

"Then we abandon docking," said CAPCOM.

"Hell, abandon the moon."

"Anybody got some ideas?" asked the first engineer.

"Some kind of trailer hitch," said Slayton.

"Houston, to hell with this 'try until you succeed' business," interrupted Shepard. "We're burning all the fuel."

"If it's dirt blocking the latches, the word here is to bump the lock free," said CAPCOM.

Shepard continued impatiently, "It's not working. We recommend putting back on the pressure suits."

At Mission Control, the first engineer said to himself, unbelieving, "A spacewalk?"

"We're in agreement up here to de-pressurize the cabin and hand guide the hard dock," said Shepard.

"Repeat, Kitty Hawk," said Kranz.

The second engineer frowned, "Too dangerous."

The third engineer surveyed the anxious room, "Any of you Ph.D.'s ever simulated a bronco bull ride?"

"I can exit the Kitty Hawk," said Shepard. "Hell, Houston, I can practically reach out the window."

Shepard blinked tightly and saw a bright light flash before his eyes in the space in front of his face. This light flash persisted in its shining presence before his eyes.

Shepard continued but with less certainty after the light flashes interrupted his train of thought. "Uh, window and grab the Lunar Module now."

"Al, we're negative on a spacewalk," said CAPCOM.

"Houston, I can just pull the two ships with my hands into hard dock," said Shepard.

"You ever walked in space before?" asked Mitchell with his hand over Shepard's live mike to Houston.

"A spacewalk, I invented it." Shepard answered without skipping a beat. He then reluctantly signaled a near miss with his fingers. "This close," said Shepard.

"Al," said Slayton to his friend.

"Deke, is that you?" asked Shepard. "What's going on down there?"

"Al, we got half a dozen guys drilling this probe in simulators. No reason to un-strap your seatbelt just yet, buddy," said Slayton.

Shepard took note of the calming tone of his friend. "Deke, I can put my nose to the window and practically touch that cone. It's that close."

"We're going do our best to get all of you to the moon and back in time for a Texas barbeque," said Slayton.

"Houston, do we have 'Go' for another bump for debris," said Roosa.

"That's a negative, Kitty Hawk. We're watching your reserve fuel, over," said CAPCOM.

Into the command room burst a team of engineers and simulation experts. The Chief Simulation Engineer announced the plan. "We think we may have a way."

"Kitty Hawk, standby," said CAPCOM.

The Chief Simulation Engineer said off radio contact with the spacecraft. "Change the procedure." The simulation engineers bumped the cone and drogue very hard, with the full force of his arms. The latches locked on the simulator.

"Kitty Hawk, the new procedure is as follows," said CAPCOM. "Come in faster."

"Faster," said Roosa.

"Roger. Ram the docking probe as deeply as possible into the cone," said CAPCOM. "If the first latches fail to engage, it's OK."

"Houston, that will require perfect aim, over," said Roosa.

"Roger, Kitty Hawk, standby," said Kranz. "Can somebody get me their docking coordinates? All lights up."

"Stu, we're going to be very steady," said CAPCOM. "Drive the Kitty Hawk hard up against the Lunar Module and hold it for long enough to grab with the twelve outer latches."

"What's the timing sequence?" said Shepard in a business-like manner

"A mechanical hand grab?" said Mitchell still unbelieving.

"More like a high speed game of chicken," said Roosa pulling back from his finger controls on the thruster. "Without the last minute wave-off."

CAPCOM spoke to the engineers in the control room. "Any of you guys got velocity numbers? Are we looking at DELTA V's over the limit."

The first engineer calmly seemed to accept the dire situation. "It either works this last try or we call the show."

"To be sure you guys are thinking of pulling us back.... revving our thrusters and ramming the only ship that can land us on the moon?" asked Mitchell.

"Houston, we're working some attitudes here," said Roosa as he reviewed some numbers on the page.

"Close enough to see, Houston," said Shepard as he raised a hand-held TV camera to the window and the Moon outside.

"We've got the LM on the television for you," said Shepard. He was still holding a thought to encourage the spacewalk idea.

On the big screen in Houston, the flight controllers squinted to see details of the dangerous maneuver.

"Roger, Kitty Hawk. We see it," said CAPCOM.

A scarred surface appears on the Lunar Module's cone.

The first engineer was able to see immediately. "There are scratches on the cone's drogue."

"Probe is entering on target," said CAPCOM.

Roosa fingered the joystick for final high-speed ramming of the Lunar Module. "Houston, requesting a fuel status," said Roosa.

The flight engineer saw the approaching red line, off-scale reading on his computer. He signaled a thumbs-down to the CAPCOM. "Bad news on fuel."

"Kitty Hawk, we'll reevaluate following your attempt to dock," said CAPCOM.

"Roger, Houston," said Roosa.

Part 6: Luck of Thirteen Plus One

The two spacecraft appeared in silhouette, flying at bullet speeds in tandem as the service module closed in on its Lunar Module counterpart.

"Houston, we're going in," said Roosa. He fingered the service module forward.

CAPCOM looked worriedly at the control room and the TV views of a closely approaching docking cone. Quietly, he said almost to himself, "Good luck, Kitty Hawk."

"Luck, hell," said Shepard off-mike while pointing camera at the incoming docking cone.

The crewmembers checked the fuel gauge and the approach distance.

"Stu?" said Shepard turning to Roosa.

Roosa turned momentarily from the delicate aim on his target. Shepard continued to watch out the window, but very coolly and slowly recommended his command. "Stu, just forget about trying to conserve fuel," said Shepard.

Mitchell looks at Roosa and Shepard disbelieving the command's finality.

"What?" said Roosa.

"This time, juice it," said Shepard.

As the Lunar Module approached glaringly close, the astronauts braced themselves for what was anticipated to be a jolting ram into a waiting spacecraft. The effect was to broad-side the only thing that could land two of these men on the moon.

"Fuel level?" said Roosa.

"Last stop for gas, five miles back," said Shepard.

"Thrusters," said Roosa inching his finger controls forward.

"Full power charge," said Shepard.

Both ship rocked from the impact.

"No rebound. We may have capture," said Mitchell.

"C'mon, grab it," said Shepard speaking to the waiting Lunar Module. "Grab the damn probe."

The click sound came and a capture light blinked on. The three men let out a collective celebration.

"Got it!" announced Roosa.

Silence filled the small Command Module as the three waited for confirmation from the electronic panels.

"We have hard dock," said Shepard quietly.

Roosa fumbled to depress his radio transmission button in excitement. "Houston, we've got a hard dock."

"Yea, verily," said CAPCOM leaning back in his chair from Mission Control.

Kranz addressed a relieved room of mission control specialists, some of whom hold the mockup models of the cone and probe apparatus. He spoke in his characteristic understated tone. "Roger that, Kitty Hawk. We confirm."

A medley of cheers and whistles from Houston was heard in the background over the spacecraft communications.

"A dirty probe lock?" said Mitchell.

"You almost missed the moon," said Roosa.

"Houston, we're ready for business up here," said Shepard.

When the cheers in Houston died down, an interrupted pause led to a conference out of microphone range.

"Kitty Hawk, please hold," said CAPCOM.

Kranz spoke in muted tones.

"We're going to wait on the green light for trans-lunar commit, Roger," said Kranz.

Mitchell and Shepard looked at each other, convinced that they have misheard the bad news at this moment for celebration.

"Houston, can you repeat?" said Shepard.

"We're hold for trans-lunar injection," said Kranz.

"Houston, can I talk to Deke?" said Shepard unbelieving.

"No worry, Alan. We just want to make sure this thing is indeed satisfactory for docking—at least for one more dock to get you home. If she doesn't hold together, we can't commit to the moon landing," said CAPCOM.

"Houston, give me Deke!" repeated Shepard more emphatically.

"Alan, we're just going to see that this last minute marriage you've arranged up there is a keeper," said Slayton as he took the microphone.

Shepard paused in respect for his good friend, Deke Slayton.

"What's the worst case, Houston? We can't dock again on the way back from the moon?" said Shepard.

"The simulation boys are just checking the numbers: fuel, all the what-ifs? Al, if we can't dock again on the return trip, we're not coming home," said Slayton.

Shepard didn't hesitate, "We can always fly tandem. Just stay undocked on the way home. Rope-tied."

Slayton cringed at the thought of sophisticated spacecraft flying a trailer hitch scenario. "Sit tight, Al," said Slayton.

"I'm telling you Deke, Ed and I can do a simple spacewalk and come back-in the Command Module," said Shepard.

Slayton looked at the surrounding engineers with a questioning expression.

"Never been simulated for the return journey," said one engineer.

"They tried it underwater as a test. Hell of a tight rope act," continued a second engineer.

"And how to get a hundred pounds of moon rocks across the rope bridge?" asked the first engineer.

Slayton wanted to interrupt the wild speculation. "Alan, not much to do with the moon rocks during a spacewalk."

"Not a problem. I throw the whole bag of rocks over my shoulder. Just point me towards the chimney, Houston," said Shepard.

"Okay we'll pulse you when it's time," said Slayton dismayed at the limited options. "Everybody sits tight."

CAPCOM interrupted the exchange. "Kitty Hawk, we have a request from the flight doctors. You're on your nineteenth hour." He rubbed his own eyes blearily. "Get some rest, they're telling you."

Part 7: Left Behind

The next day, two newspaper reporters were interviewing Louise Shepard on the golf practice tee. Her two daughters and niece were huddled around watching the firing of balls into the distance. A NASA Public Affairs Officer looked on in approval.

"You never know what fate is going to hand you," said Louise. "When NASA sent out its first astronaut acceptance letters, Alan didn't get one. Somebody had left his letter in the wrong mailbox. Only later was he the first American in space. So you never know."

The news reporter said, "You're one of the few astronaut wives who has never been to a launch, am I right?"

"For our daughters to escape the journalists actually," said Louise.

Shepard's daughter, Laura, protests. "I've never flown."

The news reporter continued, "I'm told it's a pretty sight."

Louise said, "As Alan says, 'pretty' is what works. Anyway what am I going to do for Alan, throw in a lot of emotion and make his job harder?" Her voice surprisingly bubbled up with suppressed feeling, but she tried to stem the wellspring of confused emotion.

"When he gets home safely, I'll be there to meet his ship," said Louise.

The news reporter continued his questions, "With your background, a Christian Scientist, you must believe..."

Louise struck the golf ball for a full drive and bent over to put a new golf ball on the tee. The golf tee is snapped. Louse laughed, "I don't know, Al's full of surprises."

The news reporter persisted, "But with your faith, you had no trouble with your husband's medical ...I mean, having ear surgery to fly again?"

Louise laughed, "Wouldn't matter if I did, Al was going to the moon, that was it." Louise considered the turn of events. "But I do sometimes feel that we all get up every morning, with this tremendous faith in technology."

Louise hit another shot. She checked her hand to see if a nail is broken, then adjusted her golf glove.

"...just thinking that the wheels on the car will turn properly, and that the brakes will work when we come to the next stop light," said Louise.

In the background on the golf course, a mechanical ball machine sounded in the distance and a small roving car was circling ahead picking up stray golf balls.

The newspaperman was insistent to get some kind of story, "It must be difficult to watch your husband up there. Home now for him depends on so many people you've never even met before."

"It's half a million people who make that spacecraft fly. So it's like Al is just the sheriff, but with a big town behind him," said Louise. "You know, Edgar Mitchell, Al's copilot, he worked out in New Mexico as a real-life cowboy. No kidding. At least his wife knew beforehand what she was marrying. When I meet other women and they find out I'm an astronaut's wife, they say…"

Louise paused as if to check her honesty. "But even up there, if the brakes don't work, I know that something else will."

Chapter 3

Part 1: Card Play

The astronauts were sleeping. Mitchell was conducting an experiment in extrasensory perception, ESP. When the lights in the capsule went dark, Shepard nodded off next to Roosa. Mitchell carried a small flashlight which he tried to shield from the others in the small cabin. He read a series of geometric figures from playing cards, and then mentally sent them to Olaf Jonsson in Chicago.

Roosa watched Mitchell concentrate on each card, and then move to the next in the sequence. Mitchell concentrated intensely. Finally Mitchell turned off his flashlight and went to sleep. The whole event was extremely curious to observe in a spacecraft, but passed in silence both from Mitchell and Roosa.

Shepard was unaware of Mitchell reading ESP cards. He said, "Stu, you awake?"

"It's not like my heart rate is cooperating," said Roosa.

"If the machine is broken, we're going to fix it," said Shepard.

"And the mission rules?" asked Roosa.

"The mission is the moon. If we can't dock again, you're going to hold us in tight and when my suit is fully pressurized, I'm reaching into that tunnel and grabbing. I'll pull us together, Houston or no Houston," said Shepard.

"How'd you guys sleep?" interrupted CAPCOM.

"Not a chance I'm going to count sheep over the moon up here," answered Shepard nonchalantly.

"Any instructions, Houston?" said Roosa.

Shepard waited for a more detailed response on Houston's flight plan to the moon.

CAPCOM looked around at all the faces of the engineers who had been working round the clock; he was about to give the answer that Alan Shepard had been waiting for. "Kitty Hawk, you are GO," said CAPCOM.

"Hot Damn!" said Shepard unable to contain himself.

"Hey Alan, how are Victor Poulos' ears feeling?" asked Slayton referring to Shepard's ear problem. "Is he good for the ladder?"

Confusion appeared on the faces of the Houston management team. The flight surgeon turned to face Slayton.

"Say, again, Deke. I didn't hear you," said Shepard.

He stuck his finger in his ear as if clearing it. "Actually Victor's hearing a mild dog-whistle most of the time," said Shepard. "He's learned to ignore it."

"I never met a tougher Greek guy than that Victor. Damn fine rocket man," said Slayton smiling.

"Everything is quiet, going along extremely smoothly, and we have a happy little ship here," said Shepard coolly.

"Al, I realize this is a bad time," said CAPCOM. "But we're showing a voltage drop in one of your two batteries for the ascent stage of Antares."

The first engineer examined a flashing light on his console. "We're reading three-tenths low." If the voltage drop signals a leak, then the batteries won't be able to get them off the moon.

Kranz interrupted. "We're watching the batteries. In the meantime, we suggest strongly that you enjoy the ride."

Part 2: Downhill to the Moon

As the moon approaches, it begins to fill the entire window port of the Command Module.

"The moon is out my rendezvous window right now. We're running downhill very rapidly towards it," said Mitchell.

"We've got capture orbit," said Roosa.

"Just something wild," said Shepard with the unbelief and excitement of a teenager on the way to a romantic rendezvous.

"Fantastic! You're not going to believe this," said Roosa.

"That's the most stark and desolate looking piece of country I've ever seen," said Mitchell quietly and inward-speaking.

Only Shepard noticed Mitchell's apparent distraction with the grey sphere in the window. "Troops, let's go to work," said Shepard.

Part 3: The Blue Planet

While the moon loomed larger and apparently lonelier in the crew's port window, a NASA Mission Control mews conference was beginning in the bright Houston day.

"After the failure of Apollo 13, how is that we're still fighting these docking glitches on the way to the moon?" The reporter's question seemed to punctuate the air with technological belittlement. As if only the NASA stage knew what already had been overcome just to get as far as the lunar capture orbit, the question hung unanswered for an extra few seconds.

The NASA Public Affairs spokesman dismissed the question. "We're on the doorstep of the moon. Next."

The questions persisted. "When Apollo 13 was stranded, Al Shepard said then that the chances of 13 getting back alive were nil. Next to zero. So what's the probability that Al Shepard and Fourteen now will recover from its early hang-ups?"

"Current status?" The NASA Public Affairs spokesman tried to refocus the inquiry. "We're GO for the moon landing."

Trying a different tack, the news reporter went on: "What's NASA's response to the Soviet robot, *Lunokhod I*, which is roving the moon as we speak, just 900 miles from Apollo 14's landing site?"

"No comment, but it's anybody's guess about what's the Russian word for 'tail-gating'." The actual translation is closer to not keeping your distance: '*ne soblyudat' distantsii*'.

"Any chance of a meet-up on the moon?" asked the reporter as if writing the headline in his head. "American and Russian, man and machine."

"Obviously, we're not anticipating getting in the way of a one-ton, eight wheel robot."

Part 4: Stir the Tanks

"You're 'GO' for lunar descent," said Kranz. The NASA Mission Control team had simulated the maneuver hundreds of times.

The essential steps centered on detaching the Command Module and its pilot, Stuart Roosa, from the Lunar Module which would descend to moon with Shepard and Mitchell.

"I'm dropping us to the flat end of the parabola. More fuel for landing, Houston," said Roosa.

"Roger, Kitty Hawk," said Kranz.

Roosa turns to Shepard and Mitchell in the Lunar Module below him. The two pilots are dressed in fully pressurized spacesuits, with hands poised on module controls.

"You're moving out Antares. You seem real steady. I'm going to back away from you," said Roosa.

The Lunar Module separates from the Command Module with the moon sitting huge in the background.

"Don't forget to phone home," said Roosa.

"There it is, big as life!" said Mitchell distracted by the awesome view from the now detached Lunar Module as it descends.

The pilots watch details pass by. Mitchell and Shepard blink their eyes noticeably, seeing strange flashing lights before their eyes. Mitchell reaches out to try to touch a light, but he just grabs air.

"Ed, what's going on?" asked Shepard.

The lights flash again, and Mitchell blinks. "You don't see them? Those free-floating lights have been escorting us?" asked Mitchell.

"What are you talking about?" said Shepard.

"The lights. Don't you see those flashing lights?" said Mitchell. "I've been checking my oxygen, thinking mine was going funny and giving hallucinations."

"They're bright, aren't they?" said Shepard at first pausing, then smiling in acknowledgement.

With the training of a Navy pilot, Shepard refrained from reporting anything that might keep him from flying again. Whether inner ears disorder, dizziness, or a visual hallucination, the pilot's demeanor naturally returned to the same conclusion: all systems normal.

"So it's not just me," said Mitchell. "You see them too?"

"I've been trying to blink them away since translunar injection. Hell, I'm grateful to them. At least they're playing with my eyes and not my ears," said Shepard. "I'm still trying to ignore the damn dog whistle that's playing in my head."

"Awfully quiet up there," said CAPCOM as he interrupted to the radio silence from Houston Mission Control. "You guys still with us?" he asks.

Kranz begins to open and close the celebratory cigar box, sniffing a big stogie still encased in its wrapping. He breathes deeply the smell of success.

"We're going to stir the cryogenic tanks," said Roosa. The stirring action of the previous mission had meant its doom. When Apollo 13 initiated a similar step to stir their tanks, they lost the mission. Roosa returns to confirm: "Oxygen tank 2, now stirring."

Kranz takes his hand off the cigar box cautiously as the crew relives the critical moment of the Apollo 13 disaster: the stirring of the oxygen tanks and the sparked explosion.

"Gently, boys. Like tingling a martini," said CAPCOM putting his hand to the button that takes his comment off the microphone broadcast.

Within the Lunar Module, neither Mitchell nor Shepard is engaged in the steps that led Apollo 13 to abort. It is Roosa's job to manage his ship and their job to meet him later, after their moon landing.

Shepard blinks very hard and looks out the window at the passing craters. "I have Cone Crater, Triplet, and Doublet. Star and Sunrise. Right down there," said Shepard.

"On the nose," confirmed Mitchell.

Shepard is unable to contain his excitement. "Got' em. Yep, sure do. Hoo-ha! I think we'll know them next time," said Shepard.

Mitchell is lost in a checklist. "Final pre-landing check." His eyelids nervously begin to twitch. "Time to punch in."

Part 5: Antares Final Descent

"Roger that, Antares. You're 'GO' for lunar landing dress-rehearsal. Let's show the engineers how to land this baby before we try it for the world," said Kranz.

"Fuel levels, nominal for dress rehearsal and pre-landing check," said Shepard from the cramped quarters of the Lunar

Module. Turning to Mitchell, Shepard continued: "Just like the flight simulators, Ed."

The rehearsal was a practice run for the final descent to the lunar surface.

"Let's put this in the computer," said Mitchell.

"Got it," said Shepard. Speaking to Houston, he confirmed: "Practice descent has started. The computer is beginning the practice descent.

"On the mark," said Mitchell. Suddenly turning to Shepard, he stopped the sequence. "Everything's flashing but the engines."

Noticing the monitors were not clicking in, Shepard said, "What the…?"

"Oops! Hey, we're not showing a descent sequence," said Mitchell.

Not believing his dials, Shepard continued. "Hey, Houston, our abort program has kicked in."

The moon appears very large in the background as Mitchell glances out of the Lunar Module's porthole window.

"The computer says we can't land," said Shepard.

"We'll go to manual," said Mitchell.

"Negative. The computer will initiate automatic abort. The escape engines will fire us off target," said Shepard.

"We're showing an abort signal, Antares," said Kranz trying to suppress the memory of Apollo 13 and its unlucky reversal of fortune.

"Roger, Houston, the computer is waving us off," said Mitchell. "We can't land."

"You could reach out and touch the moon," said Shepard to himself. He was momentarily wistful reflecting on a lifetime of work to reach this stage.

"The abort signal will automatically initiate rapid rendezvous with Kitty Hawk," said Kranz. "It'll be out of our hands."

Mitchell indicates that the computer has initiated abort. "We've got a simulated stage separation, Houston. Damn the computer, it just dropped off our lower stage. It thinks we've got an emergency," said Mitchell.

In Houston, the first engineer confirmed what the computer was showing. "It's right. We do have an emergency."

"Houston, the computer's trying to fly us home," said Shepard. He was coolly portraying the confidence of a lifetime's pilot training and nearly a decade manning other people's missions to space. If anyone knew the need for calm in a crisis, Alan Shepard knew how the machines and men got each other safely home.

Shepard lifts his hands from the control in a 'no-hands' gesture.

"We're just going to have to run the simulation again. Recommence landing sequence," said Kranz.

A second engineer tried to punctuate the moment: "Any of you guys practicing *lunar interruptus*"?

Shepard overheard the remark from Houston. He could see the moon growing brighter in the port window. "Damn all if I'm pulling out..," said Shepard.

"We copy, Antares," said Kranz. "Try your descent program again."

Shepard continued to follow the plan. "Roger, Houston. We, ah…" His voice breaks up as the instruments show simulation once again underway. Mitchell and Shepard stay attentive to the dials and monitors.

"We show simulated engine start. Everything came on line," said Shepard.

Mitchell confirmed the reboot, somewhat mystified with the controls in his hand. "Descent program commencing. It's starting down."

Part 6: Light the Fuse

Applause erupted in Houston following a successful simulation in anticipation of what was no longer be simulated. "Now if it just runs like this when it's time to really light the fire," said a second engineer as he crossed his fingers.

"Antares, apparently we're showing a 'Go' for full landing this time," said Kranz.

"You bet," said Shepard.

Kranz was still monitoring the simulated descent. "We show 'Go'," he said.

Suddenly the Lunar Module screens are showing abort again. "Don't say it," said Mitchell. "I know what you're …"

"Houston! Our abort program has kicked in again," said Shepard.

Kranz speaks quietly in disbelief. "Roger, Antares." He turns off his microphone. "Damn!"

The second engineer was dumbfounded at the error readings. "Man, cry wolf or what?"

Mitchell turns to Shepard. "Al, are we snake bit?"

The moon looms larger below. Abort signal is flashing red and Shepard attempts to maneuver his joystick controls.

"This is Houston. You sure someone up there doesn't have his thumb on the abort button?" said Kranz.

Mitchell and Shepard signal as if to say: 'not me. Is it you?'

"The actual abort button was hidden in a protective shield like a panic button," said Mitchell.

Shepard ignores the question. "All other systems say yes, the computer says no."

"Houston, nobody's pushed the panic button yet," said Mitchell.

"Any of you guys got an idea, we're listening," said Shepard.

"According to everything on the panels, Al, we're smack on. Everything checks out normally," said Mitchell.

"Houston, come in. What's wrong with this ship?" said Shepard with exasperation.

"Stand by, Antares," said Kranz with looks of shrugging shoulders.

Mitchell and Shepard exchange knowing glances. "Just wait?" said Mitchell.

Kranz puts his arm on the youngest flight controller, EECOM, who looks as if he's not shaving yet.

"If you don't know what to do, don't do anything," said Kranz softly, to himself. The controllers are poised like cats ready to pounce.

"OK, men. Step-by-step," said Kranz. "First. What's our fuel situation?"

"Less than three hours," confirmed the first engineer

"How many orbits?" said Kranz.

"The flight computer's not counting," said the second engineer. "On a strictly numbers basis, we've already lost the lower stage. It's telling me that we're flying home empty handed."

"That computer seems to be muttering only to itself now," said the second engineer. "And it's repeating a 'No Go'

"Definite 'No Go'."

"I have your input." said Kranz stares back silently for a moment. He delivers conversation-terminating line. "Everybody get a phone in their hands. I want the guy who can talk to that computer."

"Yes, I got a few questions for it myself," said the first engineer. "Like how to pull its plug."

"I want the guy that programmed it," said Kranz. "The guy that loaded the program. The guy that tested the program."

"Houston?" said Shepard.

"Roger, Al. What have you got?" said Kranz.

"We're looking at a short. Electrical problem, over," said Shepard.

"Repeat that, Antares," said Kranz with uncertainty.

"The computer's fritzing," said Shepard.

"Is that technical, Al or what?" said Slayton taking the mike from Kranz and speaking directly to his old friend, Alan Shepard, the first American in space, the man who made the term 'astronaut' part of history.

Part 7: Blow the Fuse

"I'm telling you, it's a short. I tap the panel with my glove, the abort signal flickers and goes away," said Shepard.

"Roger, Antares. Can you confirm how long the abort signal goes away?" said Kranz.

"Just a blink. But it's definitely a loose switch," said Shepard.

"Any of you guys ever soldered in space before?" asked Kranz speaking to the engineers.

"We just have to reprogram the computer to ignore the abort signal," said Slayton.

"What if it's a real wave-off?" asked the first engineer. "Some kind of true abort?"

"Then it shouldn't respond to a finger tap," said Slayton.

"Negative," said the second engineer. "Not enough time to reprogram."

"We're looking at under 3 hours of fuel left. Even if we could start now, it would be untangling millions of lines of computer code," said the first engineer.

"If the abort signal is disabled and the real thing starts to run into trouble...," said Kranz not completing his thought.

The first engineer interrupts cautiously, "One hell of a chance."

The second engineer continues his thought, "Without automatic abort, they couldn't option out of a tough landing."

"Gentlemen, we're a long way from crashing landing on the moon," said Kranz. "Give me some choices."

"Get the computer programmer," commands the first engineer. "It is now his baby to nurse."

"Good," said Kranz. "And get me a sequence in case they have to fly in without abort. Fly in strictly on their own."

"The pilots may not agree," said the second engineer. "There would be no exit plan.

"Then make me one," said Kranz.

"You really think a computer and Al Shepard are going to argue for long?" said the second engineer.

"They're the best. Get with it," said Slayton.

Some engineers leave the room in a huddle.

"That programmer, he's at MIT. Boston. Get me a phone line," said Kranz.

"They'll be asleep in Boston," said the first engineer.

"Then get the man some coffee," said Kranz. "The name is Donald Eyles."

Part 8: Boston Tea

A phone rings in a dark room. The lights turn on.

A phone voice asked the question: "Am I speaking with Donald Eyles?"

Eyles answers groggily, "Yes, I think so.(muted phone talk) How much time do we got?"

He throws on a coat over his pajamas.

At the front door, an Air Force car has already pulled into his driveway and they head off into the night with the moon looming in the windshield about a quarter million miles away.

On the way to MIT's Draper Lab, the driver says: "This is your problem. The on-board computer is reading a glitch, the automatic abort kicks in, and they get a panic attack."

The driver continues: "Barely enough time to fly home with little fuel. No moon. You've got about ninety minutes to work it out."

"What is Houston suggesting?" asked Eyles.

"You've got to fool the computer. Tell it in code that everything is hunky dory and that the President is just waiting for that 'all fine up here' phone call from the lunar crater."

"Without an abort option?" asked Eyles.

"Yes, the pilots will be on their own."

"That lunar terrain is no flatland. If they come in on a glancing trajectory, they'll sure as hell want to take her around for a second look."

"You're the only one talking to that computer up there."

Eyles stopped short in his thoughts. "It is a quarter million miles away and twice that distance worth of computer code."

"You might set your watch for ninety minutes. One and a half hours or we can forget the moon."

He steps on the accelerator and speeds off rapidly into the night. The Lunar Module circled the moon.

"Houston, we're preparing to go off radio contact," said Shepard.

"Roger, Antares. Prepare for communications loss. Three minutes until you're behind the dark side of the moon," said Kranz.

"The abort is still flickering, Al," said Mitchell.

"For the next 48 minutes, this light's flickering for us only. Prepare for communications blackout," said Shepard. He turns to Mitchell. "It's just us and the stars now."

At MIT Draper Lab, Donald Eyles is typing on a computer keyboard, the green monitor reflecting from his eyeglasses talking to himself.

Eyles speaks in computer code. "Do Loop Inactivated. End statement becomes Continue Only. Will you look at that? There she goes."

His driver unconsciously checks his watch. "The man needs some coffee."

Eyles continues to speak in jargon. "Recompile Source Code."

A second technician monitors the activity on a larger computer with many spinning reels of programming tape.

"You've got less than 15 minutes," said the technician.

Eyles ignores the ticking clock from the technician. "The 'If' Statement invalidated. Compile again. The decoy is set."

The driver continues to encourage Eyles as he attempts to save the moon mission. "You heard the man—He said the decoy is set. Now compile."

Tapes and computers whirl in activity. Eyles is tapping the keyboard and pushing back from the desk.

Eyles leans back and puts his hands behind his head. "Done!"

The driver can't acknowledge Eyles' speed. "That's it?"

"It's perfect! You did it," said the technician.

Eyles take off his glasses and rubs his eyes. He speaks to the driver. "Their computer's now more blind than a hair dryer."

Eyles puts back on his glasses. "If those guys have any ideas about waving off a moon landing, they'll have to open the window and start flapping their arms."

Silent pause as all parties look at each other.

The driver concludes: "Well, what are you waiting for? You don't just get to the moon by reading the answers in the back of the book."

"Let's get it up to them," said the technician.

Part 9: Fuel Check

In the Lunar Module, Shepard is now putting a piece of paper over the abort signal to cover the red light. It still flashes underneath the covering.

"Can we get a fuel check, Houston?" said Shepard.

"This is CAPCOM, Antares. The new computer program has been checked. We're sending it up to you," said Kranz.

Yellow lights begin to flicker as electronic signals begin to arrive.

"There it comes," said Mitchell.

"Ed, we've got you entering 60 numbers in sequence to the computer, over," said Kranz.

"Roger, that," said Mitchell.

"Will you look at that barber pole turning?" said Shepard.

"Antares, transmission completed," said Kranz.

"Roger, Houston. Abort option disconnected," said Mitchell.

Radio silence ensues.

Shepard looks at Mitchell. "We have it all." He turns back to the panel of dead lights. "This is your ballgame, Ed."

"Houston, commencing new descent sequence," said Mitchell.

The flicker of activity on the instrument panel intensifies.

In the Houston Mission Control, Kranz stands over his console like a mythical octopus with any snippet of useful information within his reach. "We've got a GO, gentlemen. Let's go to the moon."

The first engineer understood what had just been accomplished as unscripted and improvised design. "Somebody please make that MIT, Boston Yankee, a cup of tea."

In the Lunar Module, Mitchell reaches for the last switch, and then turns to Shepard. "If this works, everything's got to go like the book," said Mitchell.

"No surprises this time," said Shepard.

Mitchell throws the last switch. "Half the computer is now empty space. Too bad those programs don't weigh much. We could've just tossed a lot of electronic ballast overboard," said Mitchell.

"Houston, we've got it," said Shepard.

Mitchell pulls back from the controls. "It's all ours to fly, Al," said Mitchell.

"Good show, Antares," said Kranz.

Mitchell turns to the window and looks at the moon's surface "Al, we're coming up to mark our point," said Mitchell.

"Got point," said Shepard for the first time showing relief.

"Any of you guys know how much time .we had to spare?" said Kranz talking off mike in Houston.

The first engineer pauses then says with disbelief: "Fuel shows less than fifteen minutes."

Kranz lights a cigarette.

The second engineer continues the thought: "You can plan for ten years and fly for 4 days. Yet it all comes down to 15 minutes and whether one MIT computer programmer takes an extra few swipes with his toothbrush before getting down to business."

The first engineer registers the moment. "If it had been rush hour in Boston, we'd have lost the moon, you know. Maybe forever."

"Okay, one issue solved. The center pole of the tent is up. So I expect everyone in this room to finish what they're doing and go home," said Kranz.

The people in Mission Control ignore his command, showing no movement, but still bustling together and talking.

"I repeat. Everyone in this room is to finish what they're doing now and go home. Your next shift is here," said Kranz.

"Flight, we've got a problem here," said GUIDO.

Now the room goes silent.

"Talk to me," said Kranz exasperated. He lights another cigarette.

"We're still showing the Lunar Module's batteries as low," said GUIDO.

"Give me a timeline," said Kranz.

"The volts are nominal, but they shouldn't be anything more than a flashlight's worth of load on them at the moment," said GUIDO.

"Will they get us off the moon once we get there?" said Kranz.

"We're working it," said GUIDO matter-of-factly.

Shepard locks his VELCRO shoe soles into the floor and begins to concentrate on the landing protocol. "Houston, we're commencing with descent program," said Shepard.

Kranz pauses to survey the room. "Antares, you have a 'GO'," he said.

Mitchell takes his final bearings and straps in his feet. "She's nothing but a metal balloon with legs," he said.

"Time to put these little pink bodies on the line," said Shepard glancing out the window. "Look! It's the Fra Mauro."

"Hell, we can't afford the weight of a chair." Mitchell surveys the empty space behind him. "Much less an ejection seat."

"Let's plant this spider," said Shepard. He points to the window and the moon below in silhouette.

"Speed, 3,700 miles per hour. Altitude 46,000 feet," said Kranz.

"Roger, Houston. She's a hot rid," said Mitchell.

"Houston, we show 12 minutes to landing," said Shepard.

"Roger, Antares. 5 miles out," said Kranz.

"Coming on down. Just like the book says," said Mitchell confidently. He turns to Shepard and laughs, "No more snake bites, Al."

"Computer nominal," said Shepard. "No abort warning."

"Smooth. All go," said Mitchell.

"Visibility good," said Shepard.

Mitchell looks at signal panel with no lights "Uh, Al? ," said Mitchell. "Al, I'm not getting a landing radar update."

"I'll punch it through again," said Shepard.

"Antares, we're showing you as presently blind to the moon, over," said Kranz from Houston Mission Control.

"Roger that, Houston, we're blind. No radar updates," said Mitchell.

One Houston engineer leans back unbelieving: "Son-of-a-bitch."

"We're punching it again," said Shepard glancing out the window at the enlarging moon.

"C'mon radar. Get the lock on," said Mitchell ominously imploring.

"CAPCOM, can I get an update?" said Kranz off-mike to the room full of engineers.

"We're showing no radar. Not a bleep," said CAPCOM.

"I got to say, the book is going to pretty clear on this one," said Kranz.

"It's a definite 'NO GO'," said GUIDO.

The third engineer repeats his earlier summation: "Son-of-a-bitch."

"No go, from here," said CAPCOM.

A third engineer chimed in. "Without radar, you can't think of landing. It's like flying a canyon in the dark."

"Can we abort at this stage?" said Kranz.

The room is silent.

"Al, we've got nothing. The second punch up is still negative. No update," said Mitchell.

Shepard breathes a long pause as he looks helplessly over the controls. "Damn snake."

"Give me options, gentlemen," said Kranz. "This is the big show."

"We're looking at twenty thousand feet to impact," said GUIDO.

"And closing fast. Will the computer abort without radar?" asked Kranz.

"Negative," said GUIDO. "Even if they're flying blind, the computer is now disabled for automatic abort."

"Antares, you're at nineteen thousand feet," said CAPCOM.

Kranz breaks in. "This is Houston. We're not seeing… a lock on the landing radar system." His voice reveals a worried tone.

"Roger, Houston," said Shepard. The pilot's hand pushes some buttons with precision "We're on it."

Shepard reaches overhead to another power switch. "Trying to activate radar." Shepard squints forcefully with determination. "X-squared-plus-Y-squared," repeats Shepard trying to focus his mind on the task.

Mitchell stares unbelieving at Shepard. "No radar? No computer abort? No ejection? That's a 3700 mile per hour belly flop," he said.

In Houston, engineers monitor the touchdown clock. "Antares, you're at eighteen thousand feet and closing," said CAPCOM.

Kranz asks for new inputs. "Ideas?"

"We can take them as low as ten thousand feet before having to make any one-way decisions," said GUIDO.

"You're talking about less than 5 minutes," said Kranz. "To go or bring them out of there burning."

"Without radar, the lunar crater is going to look too steep for manual guidance," said CAPCOM.

"Houston, the onboard navigational system is not receiving any data," said Shepard with disbelief. "Not a flicker of life."

"Our landing radar is dead," said Mitchell.

Kranz turns off-mike to a conference of advisors. "The book is clear. You can't fly visually blind to the moon. Already they're flying with their backs to the target. Floating around like a jellyfish in their suits. Simply speaking, no navigation and you don't think about landing," said Kranz.

"Even without automatic abort, we can have manual abort," said GUIDO.

"Be sure about that…I mean, we may have to exit from under ten thousand feet," said Kranz.

"Al can fly with a window view," said CAPCOM.

The Mission Control room is silent.

"What?" asked GUIDO.

"Do I understand that you're recommending we flip them upside down?" said Kranz.

"This will not be zero-gravity. They will definitely be standing on their heads," said GUIDO.

The first engineer interrupts. "That's right. We can have them pitch over the Lunar Module before they reach ten thousand feet. Take a quick survey of the landing site."

"Upside down?" asks GUIDO. "With their backs to the landing target?"

"Absolutely. They take a long look over the surface," answered CAPCOM. "That way they can land without radar."

"No way," said Kranz. "They'll use too much fuel in the maneuver." He knew the delicate tradeoffs needed to balance a successful moon landing and the medical and engineering limits. In his experience, the key unknown factor was always the one no one could escape: time.

The second engineer looks for a solution. "GUIDO's right. If they land on empty, we could lose them."

"When Neil Armstrong touched down, he had two seconds of fuel left," said GUIDO. "And that's without anything like this, some kind of last-minute, fuel guzzling, blind fly-by of the landing site."

"Negative," said Kranz. "It's too risky."

"The simulation book is pretty clear on this one," said GUIDO. "No navigation and you're looking at a certain abort. It's not a choice, gentlemen."

Kranz frowns. "Do we know whether this is just a bad light or are we really getting no radar?"

The first engineer continues his touchdown clock. "Antares, this is Houston. We're showing 17,500 feet."

"Roger, Houston. You guys find anything?" said Shepard.

"That's a negative, Antares," said Kranz.

Mitchell looks blankly at Shepard. "Seventeen thousand," said Mitchell.

"If somebody's got any answers, we're listening," said Shepard.

Long pause as astronauts look at each other. A quarter million miles away, the Houston engineers look at each other with similar disbelief.

Kranz interrupts. "Antares…" Kranz turns off his microphone so that only those in view can hear his proclamation: "This is my responsibility."

"Roger, Houston," said Mitchell.

Kranz begins slowly and then continues with hesitation. "We should go over the procedures." Kranz paused not wanting to finish what they started. "To abort the descent."

The moon looms larger in the window. "Can you repeat, Houston?" said Shepard.

Kranz seems pained by his admission. "Fourteen, we're pushing towards an abort procedure."

The first engineer speaks to the second engineer, leaning back in his chair. "Look closely. That's the last one."

"We abort on my countdown," said Kranz. He could no longer hide the sinking feeling that Apollo 14, and likely the rest of the moon landings, would stop on his countdown.

On board the Lunar Module, Mitchell spoke in dull, lifeless and mechanical voice. "Okay, at thirteen thousand feet, we swing up and activate the ascent program."

"Affirmative, Ed," said Kranz in a defeated voice.

Shepard answered testily. "We're aware of the ground rules, Houston." He turns to Mitchell. "If we don't get to the moon, this may be the last space flight. I have no intention of being the first American in space. And also the last."

Mitchell momentarily looks out the window. "I grew up in the New Mexico desert...and Al, the moon...that's the most stark, desolate looking piece of country I've ever seen," said Mitchell.

Shepard interrupted him sharply: "What's your point?"

Part 10: Gut Check

"CONTROL, what's our fuel?" asked Kranz off-mike.

CONTROL itemized the situation. "You have less than four and half minutes. The ascent will bury your fuel needle."

"Thank you. GUIDANCE," interrupts Kranz as he addresses the navigation expert. "What are our visuals?"

"No radar. They're blind at fifteen thousand feet," said GUIDO.

"That's it, gentlemen. By the book," said Kranz. "Countdown to mission abort will commence at fourteen thousand feet," said Kranz.

Shepard answers the pertinent question to himself: "The hell it will!"

"Antares, can you repeat, over?" said Kranz.

Shepard remains silent.

"Passing fifteen thousand feet," said GUIDO.

"Ed?" Shepard pauses. "If the radar doesn't kick in, we're going to turn her over and fly her down."

Mitchell is silent.

"Dammit! We both know we can do it," said Shepard.

"That's what we came here for," affirmed Mitchell. "A perfect mission."

"Antares, passing fourteen-five," said Kranz. "Mark for my countdown to mission abort."

"Okay, Al?" asked Mitchell.

"Yes," said Shepard pulling down an additional control box.

"As long as we both know the risks," said Mitchell. "The light up here, the shadows, our depth perception...We're going to have visual disorientation. The electronic eyes are already blind, and we're going to be shadow-boxing ourselves."

Shepard narrows his eyes in preparation for landing.

"We could be on top of a crater rim before we know what hit us," said Mitchell.

"You're right, Ed," said Shepard. "Damn right. Couldn't be any worse than bringing a sick jet down at night on the deck of an aircraft carrier, could it?"

"Al, this is a mite different," said Mitchell.

"Hell, yes. It's not pitch black down there," said Shepard. "And it's not plowing a jet through the North Atlantic in forty-foot swells pitching the deck." Shepard pauses. "You're right, Ed. This is different." He turns to the controls. "It's easier."

"Antares, commence abort mission countdown," said Kranz.

Mitchell turns to take the Lunar Module controls: "Sure, Al. Piece of cake." He tightens in shoes into the VELCRO floor for landing. "Maybe a thin piece, but," said Mitchell.

"Fourteen five, Antares," said Kranz.

"We copy, Houston," said Shepard nonchalantly.

Slayton looks quizzically at the console, recognizing the determination in Shepard's voice. He speaks to the first engineer. "What do you suppose Al's up to?" said Slayton.

The first engineer acknowledges: "He's sure not listening to us anymore."

Part 11: Mutinous News

A news reporter summarizes the crisis on camera. "The word here from Houston is Apollo 14 is now encountering trouble in its lunar descent. "

The news reporter continues. "It appears as though Apollo 14's landing crew will have to jettison its lower stage and thus try to make a re-dock with its mother ship, the Kitty Hawk, now circling over their heads in lunar orbit. In the flight planner's mind, of course, this brings up the question again, as to whether the injured Lunar Module has cleared the debris from its docking collar which prevented a hard-dock three days ago. If not, the astronauts will have less than two days' worth of oxygen in the Lunar Module, which would have to attempt to ride home in tandem with its mother ship."

A second news reporter interjects: "And of course, this brings up the prospects of facing another Apollo 13."

"I think the NASA officials I've talked with are in agreement that the US space program would never recover from a second setback of that magnitude."

Part 12: The Skillful Prayer

"Any of you guys, got some good prayers, I'm willing to listen," said Kranz.

"Antares, you're two hundred feet from abort countdown initiation," said CAPCOM.

Shepard is pushing buttons maniacally. "Copy, Houston, we're trying to reactivate," he said.

"Still no navigation, Antares," said GUIDO.

"We're still seeing no apparent malfunction," said CAPCOM.

"Then give them some radar!" said Kranz in a heated voice.

Mission Control continues the touchdown clock: "Antares, one hundred feet to abort initiation."

"Fuel shows less than 3 minutes remaining," said GUIDO.

"Ed," said Shepard. His face has become intense.

"I can fly this thing like I'm wearing it," interrupted Mitchell. "Just don't make me wear it."

"I know we can bring it down," said Shepard.

"Go to admit, Al, it's going to be a first," said Mitchell.

"This is Fourteen, dammit. Let's plant this," said Shepard.

Ed Mitchell snaps on his glove strap. "Blind at 4000 miles an hour. We'll leave a hell of an impact crater."

CONTROL continues the touchdown clock. "One hundred feet to abort sequence, over."

Shepard gestures a hushed motion to Mitchell.

"Do you read, Antares?" CONTROL askes the question that is on everyone's mind. "One hundred feet to abort initiation."

Deke Slayton breaks in to Houston Control. "Damn it, Al! What are you doing?" asked Slayton.

Halfway around the world, the equivalent Russian Mission Control is watching lunar events unfold. The commander, Komski, is sipping vodka and reading a teletype page.

"Tell the Kremlin that the Americans on 14 are coming back. They've lost radar and navigation," said Komski.

His Russian assistant raises a friendly question: "Shall we contact Washington?"

Komski pauses while taking a long drink. "If they can't help themselves, what will we do?"

In Houston Mission Control, the urgency of a response from the Lunar Module forces the engineers to lean forward.

CONTROL continues his descent countdown, "Fifty feet."

"Antares, we're going to try one last time," said Kranz.

In the Lunar Module, Shepard sits silent, listening.

"Can you reset your circuit breakers?" said Kranz.

"Pull the plug at fifty feet? She may not cold start," said Mitchell.

"Houston, we copy," said Shepard.

"Hell, it works for my toaster," said Mitchell.

"Let's pull the plug," said Shepard.

Shepard pulls the circuit breaker, the computer screen goes blank, and then resets on a new power line.

"We've got you below 13,000 feet. Have you got anything yet, Antares?"

Silence and broken static fill the airwaves.

Shepard says to himself. "Damn!" He responds to the awaiting Mission Control team. "Negative, Houston. No navigation."

"That's it. Prepare for final abort," said Kranz.

Shepard nods to Mitchell that landing is going forward anyway.

"Prepare for landing descent," said Shepard.

"You sure about this, Al?" said Mitchell.

Shepard bears down on the landing controls.

In the Russian Control Room, the flight controller is reading some more pages from a teletype.

"Cancel that message," said Komski. "It's some kind of discipline problem. They've got a mutiny on 14."

His Russian assistant poses an awkward question about an international rescue attempt, "Can they reach the *Lunokhod* rover?"

"What for?" said Komski.

The Russian assistant continues, "If their batteries are low..."

"Without radar? In that rugged landing field, they'll be crippled irreparably or miracle workers," said Komski.

In the Houston Mission Control room, Shepard's action plan is slowly being understood.

"Uh, Flight, I'm not sure about the numbers, but we've got Antares with a trajectory change," said the Electrical exper in the room or EECOM.

"Abort is negative," said GUIDO. "Antares is entering its final lunar descent."

"Get me some velocities..," said Kranz.

"They're commencing with the flight plan," said EECOM.

"Can the computer not abort or what?" said Kranz.

"Got me? Ask MIT," said GUIDO.

"Then who's flying them down," said Kranz.

It begins to dawn on Kranz that the landing is going forward.

"They may just be drifting," said GUIDO.

RETRO said, "If it's the computer, then..."

"It's Al," interrupts EECOM affirmatively.

Alan Shepard's good friend and astronaut office boss, Deke Slayton, summarized the plan. "Al's not getting waved off," said Slayton.

"The son of a bitch is going to land on the moon," said GUIDO.

Silence fills the Houston Control Center.

"Antares? Abort lunar landing on my mark, ten...nine ...Antares, over?" said Kranz.

Silence from the Lunar Module.

"Communications! Have we lost anybody," said Kranz.

"Negative," said CAPCOM.

All eyes are looking up at the big screen in Mission Control Room. "They're listening," said Medical.

"Do you copy, Antares? Commence abort from the moon," said Kranz.

Disbelief begins to fall on the faces in Houston. "Hell, Al's going for the brass ring," said CAPCOM.

More silence from the moon. Static breaks through the silence. "Hold on, Houston," said Mitchell.

"We're listening, Antares," said Kranz.

Suddenly computer screens flash numbers in Houston as a data stream initiates.

Shepard and Mitchell sit in the Lunar Module with its claustrophobic confinement and a quarter million miles from anyone who could help them if anything goes wrong. As the pilot of the lunar lander, Mitchell turns to Shepard, his commander.

"Al, will you look at that," said Mitchell.

Shepard interjects as interior lights being to flash. "Houston, we've got a radar lock," said Shepard.

Part 13: The Confirmation

Silence fills the crowded room in Houston, as all activity is suspended. Numbers are streaming across computer consoles. Suddenly, all activity in the control center is buzzing. Roosa's and Mitchell's wives are seen in a viewing area behind glass. The wives begin clapping wildly.

Kranz holds a finger over one ear to quiet the noise. "Antares," said Kranz. He looks at consoles lit up triumphantly. "We're confirming incoming data on the onboard navigation system."

The first engineer spins in his chair rapidly to address Kranz, "We're up and running. All go."

"Antares, all systems are functioning. Your eyes are up," said Kranz.

Shepard coolly turns to the window. "We copy that, Houston." He looks unbelieving into space. Shepard and Mitchell fix their attention on the flight controls in their hands.

Kranz surveys the consoles and flight controllers before him. The conversation goes quickly around the room, in a roll-calling fashion.

"Gentlemen. Status. CAPCOM?" said Kranz to the Capsule Communicator responsible for reporting directly to the astronauts.

"Go, FLIGHT," said CAPCOM.

"EECOM?" said Kranz to the Electrical, Environmental, and Consumables Management Officer.

EECOM checks a console before him. "Go."

"G.N.C.?" said Kranz to the Guidance, Navigation and Control Systems Engineer who was responsible for the reaction control system and the Command Module's main engine.

G.N.C answers: "GO!"

"CONTROL?" said Kranz.

CONTROL answers, "CONTROL, Go!"

"TELMU?" said Kranz addressing the telecommunications expert in the abbreviated command language common to this particular room.

"Go, FLIGHT," said TELMU.

"RETRO?" said Kranz to the Retrofire Officer who was responsible for the engine fire that took the mission into trans-lunar orbit.

RETRO replies, "Retro, go for landing."

"FIDO?" said Kranz to the Flight Dynamics Officer responsible for certifying the spacecraft trajectory was on-course.

"GO," said FIDO.

Kranz pauses in his roll call to assess the significance reaching this critical junction.

"GUIDO?" said Kranz to the Guidance Officer.

"GO," said GUIDO.

Kranz lights a cigarette. "Antares, you are 'GO' to land."

Mitchell answers the roll-call. "You better believe, Houston."

Chapter 4

Part 1: Dream Mission

In his bedroom alone, the scientist and psychic researcher, Olaf Jonsson, stirs.

"The 'star'," said Mitchell. Fast asleep, Jonsson suddenly starts and wakes up wide-eyed from his slumber. It's as if a dream has disturbed him. He checks the clock and on Olaf's nightstand are scattered a shuffled deck of ESP cards. Jonsson blinks hard and picks a random card.

In Houston's Mission Control room, a myriad of switches are now automatically sequencing.

"What's happening?" said EECOM.

"Circuit breaker has reset. Houston, we pulled the plug. Roger that," said Shepard.

"Each flight," said CAPCOM. "It's like a new duck popping up at the shooting gallery."

Deke Slayton shook his head in disbelief. "Alright, Al. Let's go to the moon."

"Your altitude is 12,000 feet and descending," said GUIDO.

"Four square on the deck," said Slayton sitting up straight in his chair.

Shepard whoops and wildly pats Mitchell on the back. "Hell, man, we've hung it out," he said.

111

A delayed war cry of joy comes over the Houston squawk box.

Shepard fumbles to put his gloved hands back on the controls. Involuntarily his hand gives a momentary flex with fingers extended. "But never closer than this one," said Shepard.

"Ten thousand, two hundred and closing," said GUIDO.

"Houston, we're pitching over," said Mitchell.

Kranz takes a long cigarette drag. "You're looking good. Communications?" asked Kranz.

"GO," said CAPCOM.

"Guidance," said Kranz.

"GO for lunar touchdown," said GUIDO.

"Antares. Let's go," said Kranz.

The moon looks rugged and stark out the window over Mitchell's shoulder.

"Now there's a rough runway," said Mitchell. He turns to Shepard. "You're planning to wheel this down?"

"What a beauty," said Shepard, breathing again. Since being the first American in space, Alan Shepard had heard others describing the starlight and blue earth views. He had seen others land or splashdown to applause in Houston.

"Cone crater," said Mitchell. He identified features that previously had just been points on a moon map. The lunar glow was reflected off the window and Mitchell's dilating eyes. "It looks less than a jump away."

"Six thousand," said GUIDO.

"Control, what's our fuel?" asked Kranz.

"Low," noted CONTROL. "Antares, we're showing a needling fuel level."

Shepard shakes his head. "Come on, I need half a minute."

"One thousand, and we're right on the money," said Mitchell.

Shepard blinks his eyes tightly and shakes his head as if to clear away some distraction. Light flashes appear in the empty space in front of his face. "Our visitors are back," he said. Shepard's eyes were playing tricks on him as the moon loomed in front of the Lunar Module's windows.

"Light flashes?" asked Mitchell. "I got them too. Should we tell Houston?"

"Antares, from here, you are a big 'GO' for landing," said Kranz.

Shepard shakes off the momentary hallucinations and turns to military formality. "Thank you sir," said Shepard.

The Lunar Module skidded and rocked. A few boulders rise suddenly during the descent.

"Whoa, it's a quarry field," said Shepard taking note of the unexpectedly rugged terrain below him. "Shifting course. We're moving up a bit."

"Fuel nominal," said Kranz.

Mitchell suddenly points out the window. "Over there Steady, down, down, a little nudge Al, we've got a cloud of dust like you wouldn't believe…"

"Twenty feet and descending," confirmed Shepard.

"Gentle. Contact!" said Mitchell.

As the Lunar Module touches down, Shepard immediately punches the controls. "Throttle's off."

There was a long silence as Houston watched the speedy turn of events.

From their home, the Shepard family begins a slow steady clapping which the whole control room warily chimes in to join the momentary reprieve.

"Houston, uh," said Mitchell. "We're on the surface."

Part 2: Family Affairs

In Houston, the Shepard family finally breaks into laughter and tears. The living room is full of people with half-eaten dishes and overflowing ashtrays on the tables.

"Did you tell Mrs. Shepard about the low batteries on the Lunar Module?" said a NASA Public Affairs' official.

Another NASA official shakes off the question, "No," and claps his hands loudest of all in the full living room.

"Old Man Moses has found his Promised Land," said Louise Shepard, looking on with admiration.

On the lunar surface, Mitchell turns to shake Shepard's hand as the only two inhabitants of the moon. Mitchell pulls back his hand suddenly.

"The truth." said Mitchell.

Shepard's hand is rock-steady. His eyes narrow.

"Don't bother, Ed," said Shepard. "You know this thing is like backing down a freeway without a rearview mirror."

"Tell me this. You were going land this thing anyhow?" said Mitchell. "With or without navigation."

"What's your guess?" said Shepard, steely eyed.

In the Command Module orbiting the moon above the two men on the surface, Stuart Roosa had been a calm spectator to the unfolding events below. "Houston, this is Kitty Hawk. We got them. No wings. No wheels. Just a big fire in the hole," said Roosa.

"Roger," said Kranz.

"They've got the flattest landing spot in the vicinity," said Roosa. "I can see the crater in the near foreground."

"We had them figured for abort." said Kranz.

As the boss to all of the astronauts, Deke Slayton did not admonish his comrades. "Hell, they were snake-bit without first aid. Over."

"Copy that, Houston. You guys got discounts on pacemakers down there?" asked Roosa. He was to rendezvous with the top-half of the Lunar Module after Shepard and Mitchell left the moon surface. "I'm not thinking about coming home with two empty seats."

Part 3: Public Affairs

A NASA Public Affairs Officer was pacing before an audience of NASA brass and practicing his delivery with a muttering voice. The press had not entered yet. Deke Slayton enters the room briskly and takes a seat. The press rehearsal commenced with the nervous Public Affairs Officer clearing his throat and beginning to read his prepared statement.

"Give it to me," said Slayton.

A NASA Public Affairs Officer prepares to present the official timeline of events to the press corps. He begins reading "In a dramatic landing sequence..."

Slayton interrupts, "We never say the word 'landing'. There's only one real landing in this mission and that one will be in the Pacific." Slayton was less concerned about the official version of events and more concerned with jinxing a pilot, his friend, Alan Shepard, with a half-finished story. "I hope you have got something else for me."

The NASA Public Affairs Officer continues his text, "Prior to uh, landing..." He corrects himself. "I mean, touchdown."

The officer finally blurts out what summarized the last few hours on the moon and in Houston. ".An abort sequence was fully initiated on approach to the moon landing site..." He corrects himself. "I mean, not landing, uh...Fra Mauro, the Apollo 14 crew led by astronaut Alan Shepard and Edgar Mitchell, narrowly escaped..."

"Hell, son!" said Slayton. "They damn near mutinied."

The NASA Public Affairs Officer is now shaken. "...the astronauts narrowly escaped a certain disaster for NASA's Apollo program."

"Cut that. This is not one of your public relations' disasters," said Slayton. "It's a head-on collision."

"Uh, mmm." The NASA Public Affairs Officer tries to compose the press release on the fly. "Following the perilous voyage of Apollo 13..."

Slayton just looks up, correcting without words, and the room momentarily goes silent. The NASA PAO Officer is near breakdown

He begins reading more quickly now. "The crew aboard the Antares spacecraft diverted a complex computer failure and then the jammed radar which would have rendered their landing...uh, touchdown, blind."

Slayton finally shows a faint sign of approval.

"NASA officials in Houston contemplated an upside-down flyby," said the NASA Public Affairs Officer.

Slayton interrupts again. "Alan Shepard contemplated it. We just watched."

"...in which the astronauts would literally have been flying upside down and with their backs to the landing target," said the NASA Public Affairs Officer.

Slayton pauses to absorb the events. He speaks with emphasis, still unbelieving. "At supersonic speed," said Slayton.

"But in the closing moments as the abort countdown proceeded, a final command sequence to reset the on-board circuit breakers brought the astronauts back to radar navigation ...only minutes before a lunar impact," said the NASA Public Affairs Officer.

Slayton appears to be falling asleep, right in the middle of the closing remarks from the rehearsed press release. This only adds to the nervousness of the PAO officer.

"NASA officials said everything is going forward smoothly and lauded the courage and determination of the crew and its engineering team," said the NASA Public Affairs Officer.

The room is silent. Slayton appears to be dozing.

The NASA Public Affairs Officer speaks helplessly. "That's all."

Slayton keeps his head down, as if still asleep.

The NASA Public Affairs Officer. turns to his colleagues. "Sir?"

"I was just listening to your voice," Slayton looks up. "And whether or not it sounds like you're trying to bullshit me."

Part 4: Pushing Empty

"We show they were pushing empty," said Kranz. "Less than 7% fuel even from a low entry orbit."

"Will somebody please tell these pilots that there are no gas stations on the moon?" said the first engineer.

"It is close to the line. Somebody tell me how I can carry around 2500 canned emergency procedures in my head, memorized to the letter, and then Al Shepard..," said Kranz.

"You got to plan for this kind of surprise," said GUIDO.

"Yea, right," said CONTROL. "Next time we'll simulate a traffic jam in Boston and a plug that won't come out of the wall socket."

Part 5: Loopback

Shepard's daughter, Laura, sits at the dinner table with her cousin, Alice, who lives with the Shepard family. The television is on in the background.

A NASA Public Affairs Officer is eating a huge meal at the table and the two college age girls look on in amazement as he consumes plate after plate. They look somewhat frightened as a dessert of cheesecake is brought to the table. Louise Shepard is cleaning dishes at the sink.

"Nine rungs on the ladder, I counted them," said Laura. She knows that her father was derailed by dizzy spells from his career. She knows that one ladder still stands between him and the half-dollar sized satellite seen from Houston.

"...if the moon has quicksand?" said Alice. "He doesn't have any poison pills? Oh my…"

The NASA PAO cuts into the cherry topping on the cheesecake.

"Stop it," instructs Laura. "But I wonder?"

As she speaks, Shepard's daughter instinctively offers an extra napkin, which the NASA Public Affairs Officer happily tucks in his shirt like a bib.

Laura continues, "If that thing is up there going round and round the moon and...?" She cannot finish her thought. "What would I?"

The NASA Public Affairs Officer. turns to Louise Shepard. "If I were Alan, I'd sure be beating a quick loop getting back." He is satiated but picks an awkward moment to be flirtatious with a woman whose husband is now a global celebrity. "Some kind of home cooking."

Louise Shepard attempts to turn on a blender. She pushes the button two or three times, but the blades don't move. In frustration, she walks away from the blender and the blade spontaneously begins to turn noisily.

The television is turned up as they announce the start of the moonwalk.

"Girls, it's time," said Louise nervously calling her daughter and niece to the living room.

In the Lunar Module, Shepard and Mitchell prepare for their first walk on the moon.

Shepard untwists a urine tube tie from his leg. "Ed, will you look at this one?"

Mitchell immediately grasps the last medical challenge keeping them –and the rest of the world—from witnessing the first return to the Moon since the perilous Apollo 13 crisis.

"Houston, it's going to take us an extra 10 minutes here. Can you call off the cameras and TV coverage?" said Mitchell.

"Do we have a problem, Antares?" said Kranz.

Shepard proudly unties the urine tube to his space suit. It has become tangled during flight and its unknotting now cuts into their precious time on the surface.

"Negative, Houston," laughs Mitchell. "Things a bit cluttered in Al's spacesuit. Just don't tell Louise that his tubes have been tied."

"Al? Is Victor ready to take the ladder?" asks Slayton. Deke understands that there is actually another medical challenge that Shepard faces if he is actually to set foot on the moon.

Shepard looks to the hatch door in the Lunar Module. "Roger that," said Shepard.

"We're on the front porch. Preparing for first lunar excursion." Mitchell turns to Shepard as he steps out onto the ladder. "Houston, you can't imagine," said Mitchell. "The silence..."

Shepard stands at the top step. "It certainly is." He is temporarily speechless and pauses to gather his thoughts. "Here at Fra Mauro."

Shepard looks down at the ladder. Its nine rungs look like a steep drop of gold-foiled steps. Shepard steadies one arm against the door, checks his balance and Mitchell offers a hand.

"Apollo 14 on the moon," said Mitchell.

"It's been a long way, but..." Shepard looks up at the earth as he steps onto the surface of the moon. "We're here."

Part 6: Jewels

"Not bad for an old man. Even Nixon says you've got the nursing home crowd dancing," said Slayton.

A stream of geologists leave the room. Their mission is about to begin. Lee Silver, the geology professor, leads the way.

"Let's go to work," said Silver. "Get me some jewels.'

The last trail of engineers entered the backroom of geologists as they were exiting. Two of them carry candles and linen table clothes.

The first geologist is carrying a pot and hot pad. "Four days of rotation. We're geologists and we're pitching camp."

Silver opens the cooking pot carried by one of the engineers and smell-tests the contents. "Gumbo?"

"We got 200 hungry people. Your wife's?" said Slayton.

The first geologist volunteers that he was the cook. "Mine."

Silver is skeptical. "No."

The first geologist is trying to reassure that their long days will not leave anyone famished. "Family recipe."

"What?" Silver mechanically mocks the idea of a geologist cooking dinner for 200 guests. "Turn on heat. Empty beans from can. Stir."

On the moon, the Fra Mauro landscape extends to the horizon. Shepard begins to churn up dust as he walks. "Proceeding with lunar mirror assembly," said Shepard. He unpacks a folded optics and mirror system on the lunar surface.

"As you're aware, Houston," begins Shepard, aware that he is speaking to a global audience on Earth. His voice takes on a tutorial tone. "This reflecting mirror will help us predict changes in the earth's position and ultimately its climate over eons. Can we get an alignment test?" asked Shepard.

As the mirror is unfolded, a red laser light streaks forth onto the moon's surface, bounces off three internal mirrors and returns outward into space. The bright streak of light now flows as a continuous stream of radiant particles connecting to the Earth. As he walks, Shepard churns up a cloud that makes the laser light show sparkle with other particles of moon dust.

"Antares, you're connected now with Goldstone," said Kranz.

At the Laser Station back in Goldstone, California, a large laser transmitting and receiving light begins to signal the arrival of a return laser beam from the moon. The red light beam travels in 1.3 seconds to reach the precise instrument that Shepard has just unfolded a quarter million miles away.

In the Houston Mission Control, the engineering team begins a scripted procedure to enable various transmitting stations on Earth to communicate with Mitchell and Shepard.

"Hold for Madrid," said Kranz.

In Madrid, Spain, a similar bright stream of light connects the earth with the moon.

"Madrid says: *Bien*," said Kranz directing traffic like an airport control tower. He confirms that Madrid is in contact and shifts his attention to the other side of the globe in Australia. "Hold for Honeysuckle. We'll catch them on the next rotation."

In Canberra, Australia, a lizard crawls across a desert cable connecting the laser transmitting and receiving station. A bearded man sits alone by a window looking up at the moon. Suddenly a bright laser light strikes the collection panel and he picks up his radio transmitter.

In an Australian accent, he speaks into his radio, "Houston, this is Honeysuckle. We've got Antares in our laser sights."

Kranz had practiced his Australian slang for real—*fair dinkum*-- to underscore NASA's global reach but in the moment reverts to a military tone as he bridges the conversation back to the moon. "Roger, Honeysuckle. Antares, your triangulation is on."

Part 7: Once Belonged

In the Command Module, Stu Roosa is looking out the window while a song plays in the background. He circles the moon in an orbit carefully calculated to maintain terrestrial contact until he passes through the radio shadow presented by the dark side of the moon.

"Just fantastic, Houston," said Roosa.

The song verse was playing in the background:

> *Get back, get back to where you once belonged*
> *Wearing high heels and a low necked sweater*
> *Get back, get back to where you once belonged*
> *Go home.*

On the lunar surface near Fra Mauro, Shepard takes notice of the awesome laser light show. Shepard reaches in his suit pocket and drops a golf ball on the moon. "Houston, it's like a big sand trap up here," said Shepard. "Lots of lunar dust."

"We copy," said Kranz.

"Believe it or not, Houston, I can see the Lunar Module from here. Pretty rugged country," interrupts Roosa.

"Roger, Kitty Hawk, bring us back the family album," said Kranz.

"I've got the earth in one window and the moon in the other," said Roosa squinting at the Earth. "The Great Wall of China, it's there. I've got a manmade signature. Even from here."

Roosa marveled that a distant space observer could see that this tiny blue planet had inhabitants with industrial development. He clicks off three frames in a huge, window-mounted camera and then the camera jams and the picture roll begins to rewind automatically. The camera has failed and he looks at the machine with disbelief. "Negative, Houston," said Roosa.

Roosa pulls back camera ominously from the window. "We'll be flying by maps from here on out."

Roosa presses his focus closer to the window.

"Sorry to hear that, Kitty Hawk," said Kranz. He takes inventory of what tasks are happening simultaneously on three terrestrial continents, the lunar surface and the orbiting Command Module. "We advise a roll maneuver to keep your docking probe moving. There remains a chance that we can still dislodge whatever scrap metal is blocking your hard dock before return."

In Houston, the first engineer speaks off-mike to whoever is listening. "It'll be a lonely ride home without a hard-dock. Who's leaving behind two American heroes and 95 pounds of moon jewels?"

The geology professor, Lee Silver, looks on in Mission Control as Shepard surveys the lunar landscape. He sees Shepard begin an unscheduled maneuver by dropping a golf-ball on the moon.

Kranz wants to get the science portion of the mission completed so that the engineers can work remaining problems with getting the astronauts home safely. "Antares, we've got some geologists down here..," said Kranz.

Silver interrupts Kranz and frowns at what he is seeing. "What the heck?"

"Antares, we're showing some bright white rock by your foot," said Kranz.

Shepard pulls out a long rod used for collecting lunar samples. He very methodically takes out the head to a six-iron golf club and proceeds to clumsily twist the head onto a mock shaft made from another lunar collection rod.

Alan Shepard has made a golf-club on the moon.

He begins his presentation as the world watches. "Houston." Shepard waits with dramatic flair. "You might recognize what I have in my hand the handle for the contingency rock sampler. It just so happens to have a genuine six-iron on the bottom."

In the background the Earth looms as a big blue ball with polar ice caps, clouds and blue ocean. Shepard checks that the club head is locked in place.

"Looks like you're going to need more like a five-iron to clear the water, Al," said Slayton.

The medical doctors gasped at the unexpected turn. From a physician's point of view, the chance of an astronaut falling over on the Moon, tearing a spacesuit, or wrenching his back prompted a trained eye to begin their own problem-solving checklists unconsciously.

"With my left hand, I've just placed a little white pellet that's familiar to millions of Americans," said Shepard.

Kranz: slowly grins with recognition, as he is not entirely unaware of astronaut prankishness. "Al, we don't see these white moon rocks on your collection list, over?"

The whole control room suddenly captures the plan with bewilderment.

"I'm trying a sand-trap shot," said Shepard.

Shepard cannot manage a complete swing in his bulky space suit, so he tries a one-handed chop. He catches a big cloud of dust and lands the ball a 100 yards away towards the background of the Earth.

"Listen, old man, use more backswing," said Slayton. "You're aiming for an elevated green."

In California, the celebrity comedian and golfer, Bob Hope watches the golf shot on television from his house, surrounded by famous Hollywood stars. His eyes go misty. He mutters to himself softly "Just unearthly," then laughs to his guests, "Get me those fellows' agent."

On the lunar surface, Shepard is slightly out of breath, "I got more dirt than ball," he said.

Edgar Mitchell pulls up in a cart filled with scientific instruments; the cart resembles a lunar rickshaw with an astronaut pulling moon rocks back to the Lunar Module. He disappears up to his elbows in a small mound and is momentarily hidden from view except for his bobbing head above the small hill.

"Ed, you sure you're not sinking in the soft lunar soil," said Kranz.

Mitchell points vaguely in the direction of the earth. "Negative." He takes note of Shepard's audience pleasing and decides to join in. "Hey Al, that shot looked more like a slice to me."

Shepard holds up a second golf ball to his eye and its diameter completely covers the image of the earth on the horizon.

He flips the ball like a coin toss with his space-suited thumb and drops a second ball in the soft lunar soil. He gets a determined look on his face. He takes a mighty one-handed swing almost toppling over.

"Beautiful," said Shepard.

Shepard calls out to earth again, this time more confident in his swing. "There it goes. Miles and miles and miles."

"Antares, we got a team trying to churn up a lunar handicap for you," said Kranz. "You know you're looking at one-sixth usual gravity."

Shepard walks over to the American flag, straightens its wire supports, then turns to face Earth. A pregnant silence comes over the scene. Mitchell puts down the handles for the lunar rickshaw and turns to face earth. The Houston team turns silent too and puts down their head gear consoles and telephones. A stream of joyful tears fills Shepard's eyes.

Shepard's moment of silence is broken by a large explosion over his left shoulder. Mitchell is aiming a large rifle-like extender towards the moon and firing an explosive charge. Aiming at the soil below, he fires twenty-two shots in all.

Kranz picks up his console headsets again. "Antares, this is Houston. We got a good seismic reading on that last charge. Moonquake Good."

The ground ripples underneath the two astronauts in a series of small moonquakes.

"This twelve-gauge earthquake starter has a heck of a kick," said Mitchell.

Shepard regains his duty-bound attitude. He speaks as one Navy man to another as Mitchell momentarily halts firing at the Moon. "It's just a big old gun. Put your shoulder into it."

Mitchell continues to bounce along every fifteen feet and fire an explosive charge into the moon. The scene turns up a cloud of dust and literally lifts Mitchell off the surface.

Mitchell can hardly contain his enthusiasm as he settles into a repeatable method. He starts whooping and hollering like a gold prospector. Although born in Hereford, Texas, Mitchell considers his hometown Artesia, New Mexico (near Roswell). "I've branded a few cattle in my cowboy days, but this one is something like a rocket-powered pogo stick."

Part 8: Earth and Moon Started as One

From the Command Module overhead, Roosa orbits over their landing site. "Got to admit, from up here, you guys look as busy as ants who's hill has just been stepped on," said Roosa.

The background music of Roosa's country and western songs slows down to a slow pace and eventually drains away to silence as the batteries die. Roosa reaches into his pocket and replaces the batteries to his tape-recorder.

The Command Module, Kitty Hawk, begins to shake slightly and drift downward.

"We're showing Kitty Hawk with another course correction, am I right?" said Kranz.

"We're showing a big drift every orbit..," said GUIDO.

"Stuck thruster vent?" asked Kranz.

"Negative, Houston. We've got another mascon," said Roosa.

A NASA Public Affairs Officer interrupts, "Mascon?"

"Roger that, gravity from lunar mass concentration confirmed. That moon is giving a good yank," said Kranz. The Moon has an uneven interior composition and small variations in mass cause the Command Module to correct course as if drawn by a fisherman fighting a fish.

"As you know, Houston, we've got three candidate theories for the moon," begins Shepard as he explains to his audience some of the rehearsed science lectures. "The earth caught the moon, that's the capture theory. The moon tore away from the earth, that's the ripping theory. And the third, the earth and moon formed together, the twins theory," said Shepard.

As the Apollo missions progressed, their planning become more about science and less about engineering milestones.

In Houston, a series of lunar landscape pictures are coming as hard copy photographs off a crude printer. The pictures are taken from the printer, held up for a moment to determine the top and bottom of the picture, then placed piecemeal on a huge wall as an almost 360 degree, panorama of the lunar landing site. The mosaic of pictures is impressive in its size and detail. The geologist, Lee Silver, puts a push-pin holder onto the last picture completing the lunar horizon in the control room.

"Antares, down here we've got some bets on a fourth theory of how the moon came about. Congress appropriated funds for a second planet, but when the politics got sticky," said Kranz. "They had just enough left to buy a moon."

Shepard pauses from his lecture, exhilarated at the sights on the moon. "Me, I'm counting on the ripping theory, earth and moon started as one. That way, part of Tahiti's likely to be just over that crater," said Shepard.

In Mission Control, Fred Haise, the Lunar Module commander on Apollo 13, is looking with great concentration at the lunar pictures on the control video screen. He is considering his missed opportunity.

Haise begins to speak to the astronauts now on the moon. "Can you describe the lunar soil?"

"Is that you, Fred-O?" said Mitchell. Edgar knew that he was now part of an elite fraternity. There would be handful of humans who could one-day say they were actually there.

"Good to see you firmly got your foot in the door," said Haise.

"Y'all might comment that the moon face has the appearance of raindrops, of a raindrop splattered surface," said Mitchell.

Mitchell turns back to see the Lunar Module on a sharp tilt due to one landing-foot of the spacecraft in a crater. The module appears to be tilting on an incline. "That old Lunar Module looks like it's got a flat tire," said Mitchell.

Haise pauses in disbelief, "Say again, Ed?"

A Medical Officer interrupts, "We're showing Shepard's pulse rate at 150."

"Al, the doctors say your heart's a little excited. You got a jumpy signature down here," said Slayton.

Shepard replies with decided awe, "You can't imagine, Deke." He pauses to reflect. "Someday try standing a quarter million miles from home ...looking back on creation...and taking the pulse of another world."

He takes in the sight of the Earth. "Houston, we're beginning to see the Earth eclipse," said Shepard.

"Sure you don't mean lunar eclipse?" said Kranz.

"We're now reporting the first view of how the earth can block the sun, it's an earth eclipse up here," said Mitchell. "For you, I guess it's just another moonlit night."

"Hard to believe," said Kranz. As if to underscore that no terrestrial reference frame could help them anticipate the differences in an eclipse on Earth and the Moon. "Prepare for temperature drop."

"It's fantastic. In one hour, the temperature has fallen by 300 degrees," said Shepard.

In Houston, the communications engineer lowers his headset and speaks outside the boundaries of his microphone broadcasting to the world. "I've got to say, together those guys are winded enough to qualify as racehorse," said CAPCOM.

Shepard is struggling, heavy-breathing and sweating, as he climbs to a crater rim.

"Al, we've bet a case of scotch that you guys can't climb to the top of that crater," said CAPCOM.

Shepard begins marching. "Left, right, left, right." He admits it is disorienting. "The judgement of distances is all illusion up here."

"How important is that we get to the top?" asked Kranz to his team in Houston. The question was posed outside the radio transmissions sent to the Moon.

The geologist, Lee Silver, replied, "Scientifically it's the mother lode."

"Medically, we've got them out there in no man's land," said the Medical Officer. To the physician, he had seen only four men previously spend time moonwalking. Apollo 14 was planned to extend the duration doing lunar traversals. "About a mile from the module and about to drop in their tracks from exertion."

Lee Silver was not about to be trumped by a medical doctor. He knew his wish list would only be fulfilled once and the clock was ticking. "Oh, let's give the crater a look-see. It's a damn crystal ball into the ancient."

"Overruled," said Kranz. "I want those guys on top of things if we can't hard dock for the ride home."

On the Moon, these two men were not shy about evaluating their own endurance. They had practiced and rehearsed every moment of this mission. The astronauts were not immune to improvising if it meant more, but protested if it meant less. "We're less than 100 yards away," said Mitchell.

"We're here on the map," said Shepard. He turned the map sideways. "And we want to be over there."

"We're suggesting you stop short of the Cone Crater rim," said Kranz.

Mitchell and Shepard face each other, winded and sweating from their hard activity.

"Houston, can you repeat that?" said Shepard unbelieving.

"We're less than 100 yards away. We bought the car, surely we can afford the gas," said Mitchell.

"Negative, Ed. Medical is showing that you guys are exhausted," said Kranz.

"Al, this is Deke," said Slayton. He used his first name to create an intimacy that could penetrate the formality of broadcasting to over one hundred countries. He was not speaking as Deke to Al. "How's...uh, the balance problem?"

"I'll just rest on one knee for a minute," said Shepard.

"Negative," said Kranz. He was becoming impatient with negotiations.

Deke Slayton knew when Shepard needed to hear a colleagues' advice. If Slayton hadn't been to the moon, he would find someone who had. "We've got Neil Armstrong here and based on his moonwalk, he's advising that if you go down to one knee with your spacesuit, Al, you may not be able to get back up." Slayton paused to let that image sink in. "A turtle."

Shepard stops with alarm at the prospect of laying like a turtle on his back on the Moon. While the world watched, he would have fallen without the ability to get back to his feet. Whether Deke Slayton reminded him of his balance, a doctor recorded his every heart rhythm or Neil Armstrong described the spacesuit constraints, it was the picture of a lone man looking up from his back that caught Shepard's attention.

Edgar Mitchell was absorbed in setting up a lunar instrument. "It's like plowing a field all day and just as the sun starts to set, you tell us to come inside and leave all the tractors out here running without us," said Mitchell. He had a way of putting an alien landscape into a terrestrial setting.

"You guys are the record holders for 33 hours on the moon. We've got your heart rates over 150. Why not come on home, Antares?" said Kranz. It was more statement than question.

"From up here, it looks like you guys are..," said Mitchell. angry, sweating and winded. "The finks!"

Mitchell turns to the Lunar Module. Even the previously mutinous choices that Alan Shepard had undertaken would be eclipsed by a direct refusal to end a moon walk. This choice was less of a man trying to conquer a machine and more of a choice of a man trying to overcome his own limits. Shepard quickly interrupted Mitchell to restore order after the word 'finks' came out of his co-pilot's broadcast. "Roger, Houston," said Shepard.

Shepard packs up a lunar instrument and discovers the message written by the backup crew on a bumper sticker on the lunar rickshaw.

The message reads: *Watch your ass—we're right behind you. Signed Apollo 15, The First Team.*

"We're returning to earth. Goodbye from Fra Mauro and the Sea of Rains," said Mitchell.

"Long day picking cotton," said Kranz. "You guys earned some sleep."

Shepard picks up the shaft of his makeshift pole and golf club and throws it javelin-style far out of the field of view. The javelin moves just like the Apollo rocket, spinning and spiraling towards space.

Part 9: Shepard Home

Louise Shepard and her children watch on TV as the moon landing crew begins to disappear from the lunar surface.

"Look at them now," said a NASA Public Affairs Officer. "Yesterday some reporter was asking me if they were carrying suicide pills and today we're all there." He was referring to all the inhabitants of Earth except the three temporary extraterrestrials. "Everybody."

Another NASA official questions his colleague. "Cyanide? They could get pills if someone asked. These pilots used to carry them in spy planes but I like to think on the moon, it'd be easier to just to drift off... and fall asleep."

"The Sea of Rains," said Louise.

Louise speaks softly to her daughter, now 19. "When you look at the face of the Man in the Moon, you know, where his right eye begins?" Louise points to the brightly lit satellite from their backyard. "That's where your father is. Al's smack dab in the right eye of the Man in the Moon. Just sleeping like a bear."

Both Mitchell and Shepard are in hammocks in the dark of the Lunar Module. A large creaking sound breaks the silence with metallic shifting. The two men turn fitfully in their space suits.

Shepard whispers, "Are you awake?"

Mitchell whispers too from the bottom hammock. "Hell yes, I'm awake."

"Did you hear that?" asked Shepard.

"Hell yes, I heard that. Something out there is going bump in the night," said Mitchell with alarm.

The Lunar Module has one footpad planted in a crater, giving the interior a decided tilt to one side. Shepard falls out of the top hammock very gently drifting to the floor.

"Never heard anything like that before," said Mitchell.

Neither astronaut sees the characteristic light flashes before his eyes.

Shepard rubs his eyes. "I don't see our visitors," he said referring to the light hallucinations that most astronauts had experienced at one time or another.

"No light flashes for me either. I think they're must be some racket outside," said Mitchell.

A second creak makes a metallic sound outside the capsule.

"You leave the antenna in the up position?" asked Shepard.

"Hell yes, Al. She buttoned up out there," said Mitchell.

"How far did they say that damn Russian robot was supposed to be from here?" said Shepard trying to figure out how something could actually go bump in the night when they were the only ones on the Moon..

"A good 500 miles." Mitchell smiles slightly at the prospect of meeting a Russian robot face to face on the moon. "With less than a 6 mile range," said Mitchell doing the calculations needed to dismiss Shepard's speculation.

"Besides that Russian robot is little more a vacuum cleaner," said Mitchell. "We're not exactly on the main highway here."

Mitchell shifts in his hammock and an interior creak is heard this time.

Shepard still whispers. "You don't suppose this damn thing is tipping over?"

"Why are we whispering?" said Mitchell realizing the absurdity of a noisy background sound amid the desolation and airlessness.

Both men scramble and topple over each other to raise the window shade. Outside they see the Earth framed in the window. The planet is a beautiful blue-green ball of ocean and clouds surrounded by a cloak of night sky.

Shepard finally speaks uneasily. "Wouldn't that have beat all? Fly a quarter million miles, land on the moon, work up a sweat to drown Dallas and then discover...well, we've tipped over the ship like a damn turtle."

"Try explaining that one to Houston. Oh, I took a heavier dinner before sleeping on that side of the ship," said Mitchell.

"Antares?, medical is a little concerned about sleep for you guys. Your heartrates just went above 136," said Kranz.

The voice of Mission Control clears Shepard's head. "Can we get a reading on the low Lunar Module batteries?" he said.

In Mission Control, the sleep and heartrates of the astronauts were a kind of constant worry, particularly as every spare moment on the moon was battery powered and rapidly draining away. "We're checking," said Kranz.

In Houston the first engineer makes no direct answer but turns his thumb down.

"We're still showing you a third of a volt low," said Kranz.

Mitchell and Shepard exchange looks. Mitchell crosses his fingers and pushes a switch marked: Ascent Begin.

"Just give me one cold start," said Mitchell.

"We'll begin lunar ascent countdown on my mark," said Kranz. "Ten minutes."

"Let's bring this insect home," said Shepard to Mitchell.

The Command Service Module is orbiting overhead as the two men below on the moon begin preparations to join them for a ferry ride home. Country and western music is playing in the background on an open mike. Houston controllers look at each other with bewilderment.

"Anybody want to tell me who's got an open mike?" said Kranz.

"It's not coming from the moon," said CAPCOM.

The music becomes clearer with the lyrics now understandable. The song verse playing in background echoes over the miles to Earth:

> *Betty Lou's getting down tonight*
> *Betty Lou, Betty Lou*
> *Yes, it's all true.*
> *She was bad,*
> *Her momma got mad,*
> *Now she says it's alright, Betty Lou.*

"Hey Stu, you got a girl up there or something?" said Kranz.

In the Command Module, Stu Roosa is also preparing for rendezvous. "Negative. Houston, we're now entering the Sea of Rains," said Roosa.

Roosa finishes the last bite of food from a plastic bag. He turns down the music.

"The Command Module is clean. I'm even starting to enjoy this valve-and-mash cooking," said Roosa.

He turns to stow away Edgar Mitchell's set of ESP cards in a small plastic bag VELCROED to the wall.

"Just give us one more hard-dock, Kitty Hawk," said Kranz remembering the trouble that the two spaceships had trying to tie up before departing for the Moon.

"Nixon called to wish us well. He said for us to keep our noses clean on Fourteen," said Slayton. "No surprises."

"Return that," said Kranz.

"Nixon asked whether we were working 24 hours a day on this hard-dock problem," said Slayton. "And if not, why not?"

Part 10: Why Can't We?

The Lunar Module lifts off in a cloud of dust from the lunar surface.

"Goodbye, Fra Mauro," said Shepard like a sailor leaving a port.

Edgar Mitchell took a more philosophic turn on the events of the last few days. "Every time something tough comes up, we will always be able to say, 'If we can put a man on the moon, why can't we...?'"

"...fill in the blank," echoed Shepard.

"...get one more hard-dock," said Mitchell returning to the business at hand.

Roosa does housekeeping as he prepares for arrival of the landing crew. "Houston, we're 'Go' for a single orbit rendezvous," he said to Mission Control.

"Roger, Kitty Hawk," said Kranz.

Roosa observes the fiery take-off of the Lunar Module. "Take it from an old forest fire fighter, down there that's one beautiful fire in the hole," said Roosa.

"Roger that, Smokey," confirmed Kranz. "We'll see you in forty-eight minutes... when you come out from the dark side of the moon."

"Flip side," said Roosa. He was preparing to go into the moon's radio shadow. "Communications down, Houston."

"Bring it on home, Fourteen," said Kranz.

Part 11: Why Did We?

Inside the splashdown party at Louise Shepard's house, she is clutching a white handkerchief nervously. "Al, bring home that letter jacket," said Louise referring to his Annapolis letterman's jacket.

As the Command Module had docked and transferred the two moon walkers and their bags of lunar rocks, the Earth grew ever larger in the eyes of the men who had accomplished what seemed impossible. While a week or so had passed and millions of children had watched daily the unfolding timeline on TV, at school and while huddled around the dinner table, the fourth manned mission to the moon had done what the third could not: land on the moon and broadcast to the world the drama of an adventure that will be retold by their children's children.

The gathered press that covered the story represented a cross-section of die-hard space fans, cynical journalists, and perhaps for the first time since World War II, an international crew that told another country's headlines as if they were their own. At the splashdown party arranged by NASA at the Shepard home, news reporters were told by their editors to simplify the technical story and highlight the human drama.

As the headlines were printed globally, an occasional news reporter would interject some random fact to bring the enormity of the moon mission into the realm of a street tale. "The space capsule's shields should now be burning with half the heat of the sun." Whether these extremes told the story in ways that an expectant public could fathom remained a challenge when divorced from the human events as experienced by the tired but ever alert senses of those who participated directly. As one reporter counted down the clock until the three astronauts splashed into the Pacific Ocean, he remarked, "Of course, as the world waits, it'll be the astronauts who really are waiting, both for reestablishing radio contact with Houston, and also for 21 days following splashdown while they recover in quarantine."

The splashdown site was near Samoa in the South Pacific. Aboard the US aircraft carrier, New Orleans, the deck is covered with sailors awaiting the return of the astronauts after nine days of flight. Radio silence is held during reentry.

"Hello, Antares. Over?" said CAPCOM. As the ionizing metal peels off the protective shield undercoating the Apollo 14 capsule, the Earth's atmosphere blocks all communications with the astronauts braced for impact. The waiting for their first words was a familiar but tense ritual after a decade of space travel.

"Hello, Antares. Over?" repeated CAPCOM.

The silence was expected but coiled in the guts of those who were helpless to do anything if the astronauts actually needed help.

"Keep talking, CAPCOM," reassured Kranz.

There was no response from the Apollo 14 astronauts.

"How do you receive? Over?" said CAPCOM.

A crackling voice breaks the .expectant silence.

"The ocean sure looks great," said Shepard in a broken signal masked by static.

From deep underneath the ocean, a school of fish are swimming in formation. Looking up at the sun from underneath, the undersea world seems bright and inviting. It is silent, other worldly, surprisingly like the moon. .Suddenly the calm surface of the South Pacific is struck by a plunging space capsule, its heat shield sending off steam as it hits the water. Orange and white parachutes collapse like parade banners on the ocean as seen from below. A shark moves across the South Pacific floor.

"Welcome home, Apollo 14," announces Kranz finally. After being called on to make potentially life-and-death decisions for the future of these men, it now dawns on the team that these men secured the future of more missions to follow them.

Part 12: Early Morning

In quarantine quarters after the mission, Roosa and Mitchell are having breakfast. The post-mission quarantine originated to give time to evaluate whether some moon microbe might infect an entire planet but evolved into a strange but welcome debrief before the real welcome home could begin.

Shepard's breakfast is waiting uneaten on the table while medical exams are completed. The Medical Officer completes withdrawal of a needle of blood from Shepard's arm.

Shepard rubs his arm and says in an understated tone, "You guys use an unusual number of needles."

Edgar Mitchell still wears his beard as it had grown on the way back from the Moon. A medical doctor is summarizing on a clipboard the condition of the astronauts. He signs the documents with an authoritative 'O.K.'.

The medical officer singles out Alan Shepard for a diagnosis. "OK, Al, that's it, light on the breakfast. You gained a pound while on the moon. First man to ever gain weight in space," the doctor said. "Literally in weightlessness

"You imagine?" marveled Mitchell. "It's like a pound of you is now extraterrestrial."

"And Stu," continued the doctor "you're looking at losing 12 pounds in a week." No one puts together the exchange: if a pound of Shepard is extraterrestrial, then 12 pounds – the equivalent of an English stone- is now lost in space.

A NASA delivery boy drops into quarantine a case of scotch. Roosa pulls out a bottle, smiling, and checks the label approvingly. "It happens," explained Roosa. "You start sweating over these first-trippers."

Mitchell does take notice finally. "12 pounds? Really, Stu? You leave an arm up there?"

Shepard still rubs his needle mark on his forearm. "I wouldn't say anything about a lost arm. The morning papers will get you famous," cautions Shepard. He knew more than any other astronaut what the scrutiny of a press junket meant. He had been the first American in space and thus the first to ride in the parades, both big and small. Shepard remembers, "Somebody announced that my Mercury flight costs every American less than two dollars." He is taken back a decade like a film running backwards. "For a couple years afterwards, every mail delivery day, I got a bundle of two-buck checks."

While these moments were meant for comraderies and reflective story-telling, the watchful eye of scientific evaluation was never far. Roosa asked the doctor with a genuineness that masked the doctor-patient relationship. "One thing I got to tell you, doctor, the three of us, both during the in and out flight to the moon, each one of us could close our eyes and see the brightest light you can imagine. The light was so beautiful. Just floating there in front of your face. Accompanying us where ever we went," said Roosa.

The quarantine doctor lowered his glasses skeptically.

Roosa sensed that he was jeopardizing his future career if his record was appended with a diagnosis of on-board hallucinations. "Now I'm from Oklahoma and half-hillbilly myself but that's the strangest thing I ever saw. Three men sitting elbow-to-elbow, seeing bright lights every time they closed their eyes," said Roosa. "But seeing nothing when their eyes were open."

"Oh, you mean flashes?" asks the doctor. "A couple of the other crews have seen cosmic ray flashes, big atomic particles hit your optic nerve and you see the light."

"Now wait a minute!" said Shepard. He had been senior to most of the world's astronaut corps. He was unlikely to have missed any reports of unexplained events crashing into someone's optic nerve.

"Those light flashes, they're called *phosphenes*," said the doctor.

"*Phos*-what?" asked Mitchell.

"Like phosphorescent glow," said the doctor

"No way!" Shepard was hard-pressed to continue his own skepticism since he had experienced phosphenes first-hand and knew Mitchell could back up his story. "Sometimes, two of us would see the flash at the same time. Like it wasn't just coming from in here," continued Shepard pointing to his skull.

"The high energy particles," said the doctor matter-of-factly. "They not only zipped through the spaceship hull, but also through your head and then through Stu's head and then through Ed's head."

Each astronaut squirms, as if to signal pain and then recognition. "That's nasty. It's like sharing a toothbrush," said Roosa.

"All that I'm reporting," said Mitchell, "is that medically speaking, I've had a few X-rays in my day, but I never saw any light like that. They came in waves...like liquid light but bright like the sun. Right there in front of your face. I'm telling you, it was beautiful."

The doctor changes the subject, somewhat wary of documenting the astronauts' enthusiasm for what by definition seemed impossible to document or share. He shifted to conversation to the concrete news of the day, "What about those cosmonauts?"

The room goes silent.

Roosa had read about his Russian counterparts with awe and speaks with a tribute to mutual pilot courtesy. "I heard yesterday. Damn shame," said Roosa.

"Three of them," said the doctor.

"One minute they're following our Earth orbit, laughing and talking to Moscow by radio, and then...," said another doctor.

"They just lost the pressure seal during reentry. Not a damn thing anybody on the ground could do," said Mitchell.

"Moscow even assumed the guys were fine," said Roosa. He paused to grasp the reality based on the last week in his own life. "Until they opened the hatch."

Shepard is somber and speaks as if somehow left out an important piece of news that his pilots should know. "Somebody should've told me. It's like Lovell said after 13: This going to the moon, it's no bus ride," said Shepard.

Shepard picks up the newspaper as if to avoid another moment of not knowing what everyone else was discussing.

The first headline that Shepard reads takes his mind back to their perilous mission. The news coverage begins: *"Astronaut Does E.S.P. Experiment on Moon Flight."*

"Hey, Ed, did you see this?" asked Shepard. He holds up the newspaper article. "This is the funniest goddamn thing I ever saw."

Shepard begins to read the article out loud.

"They're saying you tried to mentally transmit some geometric drawings, squares and circles, in sequence to a couple of doctors here on Earth. Pure thought power," said Shepard laughing out loud.

Mitchell glances at Roosa. Mitchell pauses while drinking some orange juice.

"I did it boss. It's true," confesses Mitchell. The reports of his in-flight experiments were now known to the world. The Lunar Module pilot entrusted with landing a flimsy spider-like spaceship on the rugged lunar terrain had undertaken an experiment to test the hypothesis that extra-sensory perception might work beyond the bounds of two earthlings. Mitchell had searched his intuition on the way to the Moon to outguess chance card choices.

Shepard stares back threateningly cold to Mitchell. Shepard does not hesitate to weigh his words, "Not on this mission, you didn't." What Alan Shepard would not admit was that he had used an alias to seek medical treatment for a pilot's greatest fear, uncontrollable dizziness. Although his flight skills and willingness to challenge Houston's authority saved his mission to the Moon, his mutiny would never be reported in wider public circles. Although Shepard had smuggled on-board the world's first and only golf club to the Moon, his pranks were public relations, not the actions of a lunatic trying to guess card choices while flying to the Moon.

Shepard knew that lunacy was a term that any mission to the Moon would never be able to shake off.

Shepard stood up abruptly knocking over his plate and then slowly sat back down. A wall of small animal cages holding white mice erupted with activity as Shepard underscored the sudden uncontrolled rage he felt towards his Lunar Module pilot.

The doctor knew the rustle of laboratory mice and last minute surprises could not erase what Apollo 14 had already written into history. He coolly looked up from the three men's medical charts. "Don't spoil your breakfast, men. Three heroes. Hell, you jockeys saved the space program."

Shepard finally gathers himself and smiles widely. "Fourteen, she was a full-up mission, wasn't she?"

Roosa raises his glass of orange juice to toast the flight, his crewmates, and the doctors. "There's a horse that couldn't throw us twice."

ABOUT THE AUTHOR

DR. DAVID NOEVER graduated from Princeton University, *summa cum laude*, and Oxford University as a Rhodes' Scholar with a Ph.D. in Theoretical Physics. He worked as a NASA space scientist at the George C. Marshall Spaceflight Center in Huntsville, Alabama. He was a semi-finalist in the 1990 NASA Group 13 astronaut selection class during the space shuttle era but grew up with a fascination for the Apollo missions and the determined group of pilots, engineers and scientists who guided their generation to go to the Moon.

www.ingramcontent.com/pod-product-compliance
Lightning Source LLC
Chambersburg PA
CBHW051215170526
45166CB00005B/1905